Digital VLSI Design and Simulation with Verilog

Digital VLSI Design and Simulation with Verilog

Dr. Suman Lata Tripathi
Lovely Professional University, Phagwara, Punjab, India

Dr. Sobhit Saxena
Lovely Professional University, Phagwara, Punjab, India

Dr. Sanjeet Kumar Sinha
Lovely Professional University, Phagwara, Punjab, India

Dr. Govind Singh Patel
IIMT College of Engineering, Greater Noida, UP, India

Registered Office
John Wiley & Sons, Inc., 111 River Street, Hoboken, NJ 07030, USA

Editorial Office
9600 Garsington Road, Oxford, OX4 2DQ, UK

For details of our global editorial offices, customer services, and more information about Wiley products visit us at www.wiley.com.

Wiley also publishes its books in a variety of electronic formats and by print-on-demand. Some content that appears in standard print versions of this book may not be available in other formats.

Library of Congress Cataloging-in-Publication Data
Names: Tripathi, Suman Lata, author. | Saxena, Sobhit, author. | Sinha, Sanjeet Kumar, author. | Patel, Govind Singh, author.
Title: Digital VLSI design and simulation with Verilog / Suman Lata Tripathi, Sobhit Saxena, Sanjeet Kumar Sinha, Govind Singh Patel.
Description: Hoboken, NJ : John Wiley & Sons, 2022. | Includes bibliographical references and index.
Identifiers: LCCN 2021020790 (print) | LCCN 2021020791 (ebook) | ISBN 9781119778042 (hardback) | ISBN 9781119778066 (pdf) | ISBN 9781119778080 (epub) | ISBN 9781119778097 (ebook)
Subjects: LCSH: Integrated circuits--Very large scale integration--Design and construction. | Verilog (Computer hardware description language)
Classification: LCC TK7874.75 .T75 2022 (print) | LCC TK7874.75 (ebook) | DDC 621.39/5028553--dc23
LC record available at https://lccn.loc.gov/2021020790
LC ebook record available at https://lccn.loc.gov/2021020791

Cover image: © Raigvi/Shutterstock
Cover design by Wiley

Set in 9.5/12.5 STIXTwoText by Integra Software Services Pvt. Ltd, Pondicherry, India

10 9 8 7 6 5 4 3 2 1

Contents

Preface

Integrated circuits are now growing in importance in every electronic system that needs an efficient VLSI architecture design with low-power consumption, a compress chip area, speed, and operating frequency. The challenge for VLSI system designers is to optimize hardware-software integration for lowering the total cost of product acquisition. So, there is a demand for better technological solutions for advanced VLSI architectures that can be done through hardware description language (HDL). Verilog HDL is one of the programming languages that can provide better solutions in this new era of the VLSI industry. The prefabrication design and analysis of such advanced VLSI architecture can easily be implemented with Verilog HDL using available software tools such as Xilinx and Cadence.

This book mainly deals with the fundamental concepts of digital design along with their design verification with Verilog HDL. It will be a common source of knowledge for beginners as well as research-seeking students working in the area of VLSI design, covering fundamentals of digital design from switch level to FPGA-based implementation using hardware description language (HDL).

The book is summarized in 10 chapters. Chapters 1 and 2 describe the fundamental concepts behind digital circuit design including combinational and sequential circuit design. Chapters 3 to 8 focus on sequential and combinational circuit design using Verilog HDL at different levels of abstraction in Verilog coding. Chapter 9 includes implementation of any logic function using a programmable logic device such as PLD, CPLD, FPGA, etc. Chapter 10 covers a few real-time examples of digital circuit design using Verilog. Chapter 11 focuses on System Verilog, distinct features, computing Verilog and System Verilog with design example.

About the Authors

Dr. Suman Lata Tripathi completed her PhD in the area of microelectronics and VLSI from Motilal Nehru National Institute of Technology, Allahabad. She obtained her M.Tech in Electronics Engineering from Uttar Pradesh Technical University, Lucknow and B.Tech in Electrical Engineering from Purvanchal University, Jaunpur. She is associated with Lovely Professional University as a Professor with more than 17 years of experience in academics. She has published more than 55 research papers in refereed IEEE, Springer, and IOP science journals and conferences. She has organized several workshops, summer internships, and expert lectures for students. She has worked as a session chair, conference steering committee member, editorial board member, and reviewer in international/national IEEE/Springer Journal and conferences. She received the "Research Excellence Award" in 2019 at Lovely Professional University. She received the best paper at IEEE ICICS-2018. She has edited more than 12 books/1 book series in different areas of electronics and electrical engineering. She is associated with editing work for top publishers including Elsevier, CRC, Taylor and Francis, Wiley-IEEE, SP Wiley, Nova Science, and Apple Academic Press. She also works as series editor for, "Smart Engineering Systems" CRC Press, Taylor and Francis. She is associated as a senior member IEEE, Fellow IETE and Life member of ISC and is continuously involved in different professional activities along with her academic work. Her area of expertise includes microelectronics device modeling and characterization, low-power VLSI circuit design, VLSI design of testing, and advance FET design for IoT, Embedded System Design and biomedical applications etc.

Dr. Sobhit Saxena completed his PhD at "IIT Roorkee" from the Metallurgical & Materials Engineering Department. He has designed a new hybrid system of Li-ion battery and supercapacitor for energy storage applications. He worked as SEM (Scanning Electron Microscopy) operator for 4 years under the MHRD fellowship. He obtained his M.Tech in VLSI from Shobhit University, Meerut and B.E. from MIT Moradabad in Electronics and Communication Engineering. He has vast teaching experience of more than 10 years in various colleges and universities. Currently, he is working as Associate Professor in the School of Electronics and Electrical Engineering, Lovely Professional University. He has been awarded the "Perfect Award" four times consecutively (2007–2010) for achieving 100% results. He has published around 12 research papers in SCI-indexed journals and reputed international conferences held at different IITs. His area of expertise includes nanomaterial synthesis and characterization, electrochemical analysis and modeling and simulation of CNT based interconnects for VLSI circuits.

Dr. Sanjeet Kumar Sinha completed a doctoral program in the Department of Electrical Engineering of the National Institute of Technology (NIT) Silchar. He obtained his M.Tech in Microelectronics and VLSI Design from NIT Silchar. He is associated with Lovely Professional University as an Associate Professor with more than 10 years of experience in academia. Over the years, he has developed an innovative approach to teaching and conducting research with undergraduates through creating and presenting course materials in both laboratory and classroom settings. He has published around 35 research papers in refereed journals/conferences including Elsevier, IEEE transaction, etc. He has published book chapters in Elsevier, Taylor & Francis, etc. His area of expertise includes microelectronics device modeling and characterization, low-power VLSI circuit design, VLSI design of testing, fabrication & characterization of CNT FET, etc.

Dr. Govind Singh Patel received his Masters degree in Instrumentation & Control Engineering from MD University, Rohtak, India. He has studied and been awarded a PhD in Electronics and Communication Engineering from Thapar University, Patiala, India. He is working as Professor in the Department of Electronics and Communication Engineering, IIMT Engineering College, Greater Noida, UP, India. He has published more than 62 papers in National and International Journals. He has also filed four Indian patents. His area of expertise includes VLSI signal processing, communication systems, low-power VLSI circuit design, VLSI design &testing and advance electronics, design for IoT, and agriculture applications. His one book titled "Smart Agriculture: Deep Learning, Machine Learning and IoT" is in the process of publication with CRC Taylor and Francis.

1

Combinational Circuit Design

This chapter describes the combinational logic circuits design and their implementation with logic gates, multiplexers, decoders, etc. Combinational circuits are the major block of any digital design or function [1]. So, a detailed overview before the design and analysis of digital circuit with Verilog modules, plays a significant role in hardware optimization to achieve the desired outcomes.

1.1 Logic Gates

Logic gates are very useful when performing a few basic operations in any digital computer system. These logic gates perform many operations in complex circuits and other control systems, e.g., basic operations like AND, OR, and NOT. The functionality of each basic gate as well as the extended version are discussed in this chapter.

AND operation:
It performs the AND operation. The circuit diagram of the N input AND operation is shown in Figure 1.1.

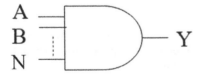

Figure 1.1 Symbol of an AND gate.

Digital VLSI Design and Simulation with Verilog, First Edition. Suman Lata Tripathi, Sobhit Saxena, Sanjeet Kumar Sinha, and Govind Singh Patel.
© 2022 John Wiley & Sons Ltd. Published 2022 by John Wiley & Sons Ltd.

The AND gate may have N number of inputs and one output. If the number of inputs are N then $N \geq 2$ conditions must be applied for input operation. Digital inputs are applied in terms of A, B, C......N, and the output is Y.

The mathematic equation is given below:

$$Y = A \text{ AND } B \text{ AND } C \text{ AND } D..........\text{AND } N$$
$$= A.B.C.D.......N$$
$$= ABCD.......N$$

The truth table for an AND gate is provided in Table 1.1

Table 1.1   T. Table of AND gate.

I/P		O/P
A	B	Y
0	0	0
0	1	0
1	0	0
1	1	1

OR operation:

It performs OR operation. The symbol for an OR operation is shown in Figure 1.2.

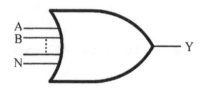

Figure 1.2   Symbol for an OR gate.

The OR gate may have N number of inputs and one output. If the number of inputs are N then $N \geq 2$ conditions must be applied for input operation. Digital inputs are applied in terms of A, B, C......N, and the output is Y.

The mathematic equation is given below:

$$Y = A \text{ OR } B \text{ OR } C \text{ OR } D.......... \text{ OR } N$$
$$= A+B+C+D......+ N$$

The truth table for an OR gate is provided in Table 1.2.

Table 1.2 Truth table of an OR gate.

I/P		O/P
A	B	Y
0	0	0
0	1	1
1	0	1
1	1	1

NOT operation:

This is also called an inverter. The symbol for the NOT gate is shown in Figure 1.3. It has a single input device and it generates an inverted output. Table 1.3 describes the truth table of a NOT gate. The mathematical equation is written as:

$$Y = NOT\ A$$
$$= \overline{A}$$

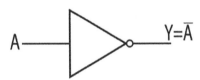

Figure 1.3 Symbol for a NOT gate.

Table 1.3 Truth table of a NOT gate.

I/P	O/P
A	Y
0	1
1	0

1.1.1 Universal Gate Operation

Universal gates are those in which any logical expression can be realized. The NAND and NOR gates are very popular and are widely used for realization of logical expressions. Therefore, these two NAND and NOR gates are used to implement other gates so these are called universal gates.

NAND operation

This is a universal gate. The operation NOT-AND is known as a NAND operation. It has N number of inputs and one output like other basic gates. However, two inputs and one output NAND gate are shown in Figure 1.4. Table 1.4 provides output values of a NAND gate in terms of inputs. The Boolean equation is given below:

$$Y = \overline{A.B}$$

Figure 1.4 Symbol for a NAND gate.

Table 1.4 Truth table of a NAND gate.

I/P		O/P
A	B	Y
0	0	1
0	1	1
1	0	1
1	1	0

NOR operation

This is a universal gate. The operation NOT-OR is known as a NOR operation. It has N number of inputs and one output similar to basic gates. The symbol diagram of two inputs and one output is shown in Figure 1.5. Table 1.5 gives output values of a NOR gate in terms of inputs. The Boolean equation is given below:

$$Y = \overline{A + B}$$

A ────────╮
          │───────▷ Y
B ────────╯

Figure 1.5 Symbol for a NOR gate.

Table 1.5   Truth table of a NOR gate.

I/P		O/P
A	B	Y
0	0	1
0	1	0
1	0	0
1	1	0

EX-OR Operation

The operation EX-OR is used in many applications. It has N number of inputs and one output like other basic gates. The symbol diagram of two I/P and one O/P is shown in Figure 1.6. Table 1.6 provides the output values of an EX-OR gate in terms of inputs. Its mathematic equation is given below:

$$Y = \overline{A \oplus B}$$

Figure 1.6   Symbol for a NAND gate.

Table 1.6   Truth table of a NAND gate.

I/P		O/P
A	B	Y
0	0	0
0	1	1
1	0	1
1	1	0

1.1.2   Combinational Logic Circuits

This type of circuit depends upon the I/Ps in that particular instant of time. A memory element is not available. A combinational circuit may have a number of sub-systems as shown in Figure 1.7.

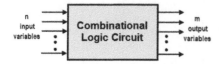

Figure 1.7 Diagram of a combinational logic circuit.

There are many ways to design these combinational logic circuits. These include:

1. Boolean expression
2. Set of statement
3. Truth table

These designs are used to design combinational logic circuits. However, a number of methods are also available to simplify Boolean function. These include:

a) Algebraic method
b) K-map method
c) Variable entered method
d) Tabulation method

Standard representation for logical functions
Any logical functions can be represented in terms of their logical variables. Logical variables and their functions are in binary form. There are two standard forms generally being used in circuit designing.

1. Sum of product (SOP)
2. Product of sum (POS)

Apart from the form above, other forms are also available to design circuits. However, these forms are conveniently suitable for the design process. This is discussed in more detail in the next subsection.

1.2 Combinational Logic Circuits Using MSI

This subsection describes the simplification and realization of the combinational logic circuits using gates. These methods are used to integrate complex functions in the form of IC. There are many devices are available such as adders, multiplexers, de-multiplexers, decoders, and multipliers.

1.2.1 Adders

An adder is a combinational logic circuit that performs arithmetic sums of binary numbers and produces corresponding outputs.

Half Adder

This is a basic adder that performs arithmetic sums of two inputs and gives the corresponding output in terms of sum and carry. The diagram of a H. adder is shown in Figure 1.8.

Figure 1.8 Block diagram of a H. adder.

A and B are I/Ps and O/Ps and are the sum and carry of the H. adder. The truth table is given in Table 1.7.

Table 1.7 Truth table of a half adder.

A	B	SUM	CARRY
0	0	0	0
0	1	1	0
1	0	1	0
1	1	0	1

Mathematical expressions for the H. adder are:

$$\text{Sum} = \bar{A}B + A\bar{B}$$
$$= A \oplus B$$
$$\text{Carry} = AB$$

The circuit diagram of the H. adder is shown in Figure 1.9.

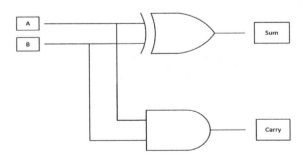

Figure 1.9 Circuit diagram of a half adder.

Full Adder

This performs the arithmetic sum of three inputs and gives the corresponding two outputs in terms of sum and carry. A block diagram of the full adder is shown in Figure 1.10. Table 1.8 provides the truth table of a full adder circuit where output variables (Sum, Cout) are expressed in terms of input values.

Figure 1.10 Block diagram of a full adder.

Table 1.8 Truth table of a full adder.

A	B	Cin	Sum	Cout
0	0	0	0	0
0	0	1	1	0
0	1	0	1	0
0	1	1	0	1
1	0	0	1	0
1	0	1	0	1
1	1	0	0	1
1	1	1	1	1

Boolean expressions for the F. adder are:
K-Map for Sum:

	B'Ci'n	B'Cin	BCin	BCin'
A'		①		①
A	①		①	

$$Sum = A \oplus B \oplus Cin$$

K-Map for Cout:

	B'Ci'n	B'Cin	BCin	BCin'
A'			1	
A		1	1	1

$$Cout = AB + BCin + Cin\,A$$

Logic diagram of a full adder

A logical representation of the F. adder is shown in Figure 1.11.

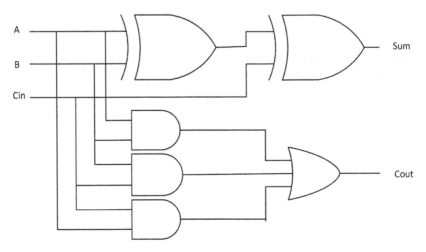

Figure 1.11 Full adder logic block.

Half Subtractor

A half subtractor is a combinational logic circuit that performs the arithmetic difference between two inputs and provides the corresponding output in terms of difference and borrows as shown in Figure 1.12. Table 1.9 provides output variables (Difference, Borrow) of the half subtractor in terms of inputs (A,B).

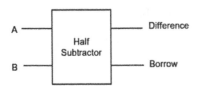

Figure 1.12 Half subtractor.

Table 1.9 Truth table of the H. subtractor.

A	B	Difference	Borrow
0	0	0	0
0	1	1	1
1	0	1	0
1	1	0	0

Boolean expressions for the H. adder are:

$$\text{Diff} = \overline{A}B + A\overline{B}$$
$$= A \oplus B$$
$$\text{Borrow} = \overline{A}B$$

Logic diagram of a H. Subtractor

The logical representation of a H. adder is shown in Figure 1.13.

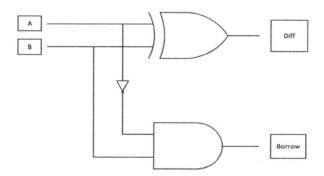

Figure 1.13 Half subtractor logic block.

Full Subtractor

This performs the arithmetic difference of three inputs and gives corresponding two outputs in terms of Diff. and Borrow. A block diagram of the F. subtractor is shown in Figure 1.14. Table 1.10 provides output variables (Difference, Borrow) of the full subtractor in terms of inputs (A,B).

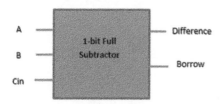

Figure 1.14 Block diagram of the full subtractor.

K-Map for Difference:

	B'Bin'	B'Bin	BBin	BBin'
A'		①		①
A	①		①	

Table 1.10 Truth table of the full subtractor.

A	B	Bin	Diff	Borrow
0	0	0	0	0
0	0	1	1	1
0	1	0	1	1
0	1	1	0	1
1	0	0	1	0
1	0	1	0	0
1	1	0	0	0
1	1	1	1	1

A mathematical equation for the F. subtractor is:

$$\text{Diff} = A \oplus B \oplus \text{Bin}$$

K-Map for Borrow:

	B'Bin'	B'Bin	BBin	BBin'
A'		1	1	1
A			1	

$$\text{Borrow} = \bar{A}B + \bar{A}B_{in} + B\bar{B}_{in}$$

The logic diagram of the F. subtractor is shown in Figure 1.15.

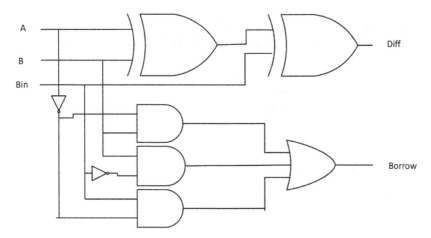

Figure 1.15 Full subtractor logic block.

1.2.2 Multiplexers

This is a type of combinational circuit. It has multiple inputs and a single output and its output depends upon the select lines. Select lines control the processing of the multiplexer which means that whatever the input is will be the output depending on select lines. If N select lines then input lines will be 2^N as shown in Figure 1.16.

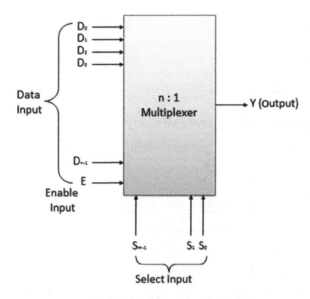

Figure 1.16 Block diagram of the multiplexer.

For simplicity of the expression, 4:1 MUX has been explained in the truth table. The truth table for 4:1 MUX is given in Table 1.11.

Table 1.11 Truth table of the 4:1 MUX.

S_1	S_0	Y(Output)
0	0	D_0
0	1	D_1
1	0	D_2
1	1	D_3

The Boolean expression for Y can be expressed as:

$$Y = \overline{S_1}\,\overline{S_0}D_0 + \overline{S_1}S_0D_1 + S_1\overline{S_0}D_2 + S_1S_0D_3$$

The logic diagram of a multiplexer is shown in Figure 1.17.

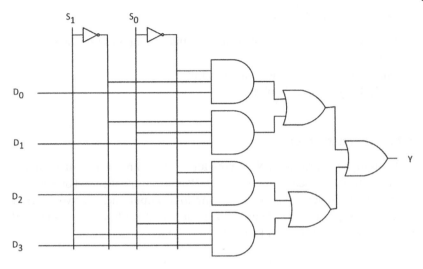

Figure 1.17   Logic diagram of the multiplexer.

Example:
Implement the expression $F(A,B,C) = \sum m(0,2,6,7)$ using a multiplexer.

Solution:
There are four variables; therefore, it needs four select lines for the process. And there are four Minterms which means these inputs are connected with logic 1 and the remaining with logic 0. For this implementation, 8:1 MUX is required. The implementation is shown in Figure 1.18.

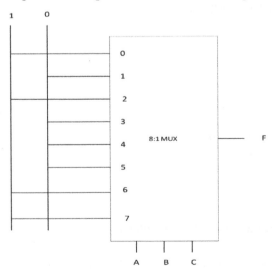

Figure 1.18   Implementation of function.

Advantages of multiplexers:

1. Logic circuits can be implemented without simplification of the logic expression.
2. This reduces the integrated circuit count.
3. Logic design can be simplified.

1.2.3 De-multiplexer

A de-multiplexer provides 2N outputs for N select lines and a single input. These controls are used to select which O/P line to route the I/P. For example, a 1 × 4 de-multiplexer has one input and four output lines for two select or control lines, as shown in Figure 1.19. Table 1.12 describes the truth table of the de-multiplexer shown in Figure 1.19.

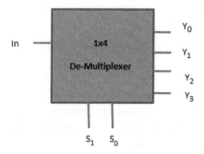

Figure 1.19 Block diagram of the de-multiplexer.

Table 1.12 Truth table of a 1 × 4 de-multiplexer.

S_1	S_0	Y_3	Y_2	Y_1	Y_0
0	0	0	0	0	In
0	1	0	0	In	0
1	0	0	In	0	0
1	1	In	0	0	0

With the help of the truth table, the de-multiplexer output can be expressed as:

A logical representation of the de-multiplexer is shown in Figure 1.20.

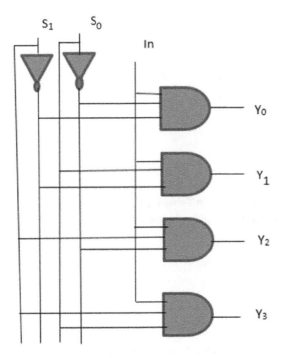

Figure 1.20 1 × 4 de-multiplexer using logic gates.

1.2.4 Decoders

This is a type of combinational circuit used to decode binary data. There are several decoders such as the BCD to seven-segment decoder, Decimal to BCD decoder etc. Here, the basic decoder 2:4 is being discussed, as shown in Figure 1.21. Table 1.13 shows the truth table of a 2:4 line decoder.

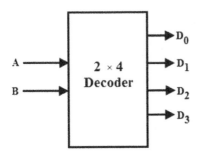

Figure 1.21 Block diagram of a 2 × 4 decoder.

Table 1.13 Truth table of decoder 2 × 4.

A	B	D_0	D_1	D_2	D_3
0	0	1	0	0	0
0	1	0	1	0	0
1	0	0	0	1	0
1	1	0	0	0	1

The truth table of the decoder is given in Table 1.10.
The Boolean function of the output is given below as:

$$D_0 = \bar{A}\bar{B}$$
$$D_1 = \bar{A}B$$
$$D_2 = A\bar{B}$$
$$D_3 = AB$$

Logic Diagram of Decoder

A logic diagram of the decoder 2 × 4 is shown in Figure 1.22.

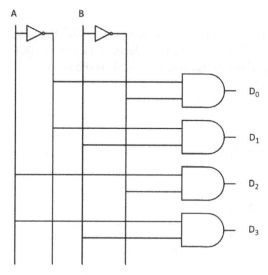

Figure 1.22 Logic diagram of a 2 × 4 decoder.

Example:

Implementation of the given functions using a 3:8 decoder.

$$F_1 = \sum m(0,3,4), \; F_2 = \sum m(2,5,6,7).$$

Solution:

Implementation of the above given function is shown in Figure 1.23.

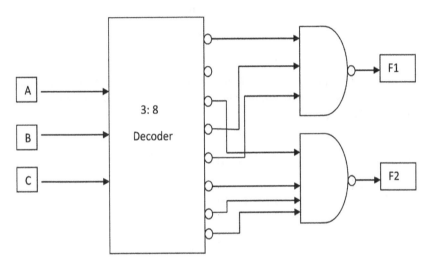

Figure 1.23 Implementation of functions using the decoder.

1.2.5 Multiplier

A multiplier is a type of combinational circuit which is used to multiply binary numbers. There are several multipliers such as the 2-bit multiplier, 4-bit multiplier, etc.

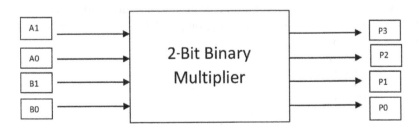

Figure 1.24 Block diagram of a 2-bit binary multiplier.

A block diagram of the 2-bit binary multiplier is shown in Figure 1.24. It will abide by the following sequence in binary multiplication.

$$
\begin{array}{cc}
A1 & A0 \\
B1 & B0 \\
\hline
A1B0 & A0B0 \\
A1B1 & A0B1 \\
\hline
\end{array}
$$

$$P3 \quad P2 \quad P1 \quad P0$$

Where P0 is the product of two bits only and P1 and P2 will be worked as H. adders, P3 is the carry output. The logic diagram of the 2-bit multiplier is shown in Figure 1.25.

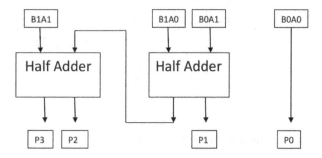

Figure 1.25   Circuit diagram of a 2-bit multiplier.

1.2.6   Comparators

Comparators are the example combinational circuit and can be implemented through a logic gate. Comparators are part of many digital and analog circuits such as ADC to DAC converters, etc. The design of comparators is usually done with the help of a truth table as shown in Table 1.14 for a 2-bit comparator. A 2-bit comparator with input A, B and output Y1(A < B), Y2(A = B) and Y3(A > B) is shown in Figure 1.26.

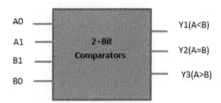

Figure 1.26   2-bit comparator block.

Table 1.14 Truth table of a 2-bit comparator.

I/P				O/P		
A1	A0	B1	B0	Y1(A < B)	Y2(A = B)	Y3(A > B)
0	0	0	0	0	1	0
0	0	0	1	1	0	0
0	0	1	0	1	0	0
0	0	1	1	1	0	0
0	1	0	0	0	0	1
0	1	0	1	0	1	0
0	1	1	0	1	0	0
0	1	1	1	1	0	0
1	0	0	0	0	0	1
1	0	0	1	0	0	1
1	0	1	0	0	1	0
1	0	1	1	1	0	0
1	1	0	0	0	0	1
1	1	0	1	0	0	1
1	1	1	0	0	0	1
1	1	1	1	0	1	0

We can obtain the out expression for Y1, Y2, and Y3 either by direct implementation with the truth table or by applying K-map reduction techniques.

$$Y1(A < B) = \overline{A1}B1 + \overline{A0}\,\overline{A0}\,B0 + \overline{A0}\,B1B0$$

$$Y2(A = B) = \overline{A1}\,\overline{A0}\,\overline{B1}\,\overline{B0} + \overline{A1}\,A0\,\overline{B1}B0 + A1\,\overline{A0}B1\overline{B0} + A1A0B1B0$$

$$Y3(A > B) = A1\overline{A0}\overline{B1}\,\overline{B0} + A1\overline{B1} + \overline{A1}\,A0\overline{B0}$$

With the above Boolean expressions, a 2-bit comparator can be implemented using logic gates such as AND, OR, and NOT gate.

1.2.7 Code Converters

There is a different format of presentation for digital data like Binary, Octal, Hexadecimal, Gray code, etc. As per the requirement, any one format can be changed into another format with the help of code converters. Several examples would be decimal-to-binary, binary-to-octal or binary-to-hexadecimal, and

Table 1.15 Octal to Binary converter.

Input in Octal								Output in Binary		
I0	I1	I2	I3	I4	I5	I6	I7	B2	B1	B0
1	0	0	0	0	0	0	0	0	0	0
0	1	0	0	0	0	0	0	0	0	1
0	0	1	0	0	0	0	0	0	1	0
0	0	0	1	0	0	0	0	0	1	1
0	0	0	0	1	0	0	0	1	0	0
0	0	0	0	0	1	0	0	1	0	1
0	0	0	0	0	0	1	0	1	1	0
0	0	0	0	0	0	0	1	1	1	1

binary-to-gray and vice versa. The design of code converters is done using logic gates with the help of a truth table (Table 1.15).

Example:

Binary-to-Octal

From Table 1.15, the binary O/P can be derived in terms of I/P as an octal value.

$$B2 = I4 + I5 + I6 + I7$$
$$B1 = I2 + I3 + I6 + I7$$
$$B0 = I1 + I3 + I5 + I7$$

Therefore, octal-to-binary implementation required the OR gate to obtain binary output in terms of octal as input variables.

1.2.8 Decimal to BCD Encoder

This is a type of priority encoder used to convert decimal to BCD number. It takes inputs in the form of numerals between 0 to 9 and gives corresponding BCD output. It works like switches; the response of the 1 or 0 turning them ON or OFF switch mode. Table 1.16 gives output values (BCD) in terms of input values.

From Table 1.16, the BCD O/P can be represented in terms of I/P as a decimal value.

$$Y3 = D8 + D9$$
$$Y2 = D4 + D5 + D6 + D7$$
$$Y1 = D2 + D3 + D6$$
$$Y0 = D1 + D3 + D5 + D7$$

Table 1.16 Truth table of a decimal to BCD encoder.

Input										Output			
D9	D8	D7	D6	D5	D4	D3	D2	D1	D0	Y3	Y2	Y1	Y0
0	0	0	0	0	0	0	0	0	1	0	0	0	0
0	0	0	0	0	0	0	0	1	0	0	0	0	1
0	0	0	0	0	0	0	1	0	0	0	0	1	0
0	0	0	0	0	0	1	0	0	0	0	0	1	1
0	0	0	0	0	1	0	0	0	0	0	1	0	0
0	0	0	0	1	0	0	0	0	0	0	1	0	1
0	0	0	1	0	0	0	0	0	0	0	1	1	0
0	0	1	0	0	0	0	0	0	0	0	1	0	0
0	1	0	0	0	0	0	0	0	0	1	0	0	0
1	0	0	0	0	0	0	0	0	0	1	0	0	1

Therefore, decimal to BCD encoder implementation requires the OR gate to obtain BCD output in terms of decimal as input variables.

Review Questions

Q1 Design and implement the given function using logic gates at gate level model:

$$F = \bar{A}BC + A\bar{B} + CD$$

Q2 Implement the following functions using NAND only.

$$F(a,b,c) = \sum m(1,2,3)$$

Q3 Implement the following multiple output function using a suitable decoder.

$$F_1(A,B,C,D) = \sum m(0,4,7,10)$$
$$F_2(A,B,C) = \sum m(1,5,6)$$

Q4 Implement the following multiple output function using a suitable multiplexer.

$$F(A,B,C) = \sum m(1,2,6,7)$$

Q5 Implement the given function using 4×1 multiplexer:

$$F_1 = \sum m(1,3,4,5)$$

Multiple Choice Questions

Q1 Select the correct representation of given function

$$f(A,B,C) = \sum 1,2,4,5,7$$

A A'B'C + A'BC' + A'B'C' + AB'C + ABC

B A'B'C + A'BC' + AB'C' + AB'C + ABC

C A'B'C + A'BC' + AB'C' + AB'C + ABC'

D A'B' + A'BC' + AB'C' + AB'C + ABC

Q2 Select the correct representation of given function

$$f(A,B,C) = \prod(1,2,4,5)$$

A A'B'C' + A'BC + ABC' + ABC

B A'B'C + A'BC' + AB'C' + ABC

C A'B'C + A'BC' + AB'C' + ABC

D AB' + A'BC' + AB'C' + ABC

Q3 Select the option which does not match with

$$f(A,B,C) = \sum 1,2,4,6$$

A $F = \prod M(0,3,5,7)$

B F = a'b'c + a'bc' + ab'c' + abc'

C F = a'bc + a'bc' + ab'c' + abc'

D Option a and b

Q4 How many 4×1 multiplexers are required to implement 16×1 multiplexer?

A 4

B 6

C 5

D 6

Q5 The output of a logic gate is 1 when all its input are at logic 0, gate is
   A NAND, EXOR
   B OR, EXNOR
   C AND, EXOR
   D NOR, EXNOR

Reference

[1] Morris, M.M., and Ciletti, M.D. (2018). *Digital Design*. Upper Saddle River, NJ: Pearson.

2

Sequential Circuit Design

This chapter describes flip-flops (F/F), registers, counters, and the finite state machine. F/Fs are part of all sequential circuit designs [1]. A sequential logic circuit can be designed using flip-flop and combinational logic circuits. Registers and counters are commonly used in digital systems. The designs of all flip-flops are described in the next subsection.

2.1 Flip-flops (F/F)

F/Fs are a single-bit binary storage device, that have two states, one is pre-set and the other is clear state. It has two outputs and these are complementary to each other. Each flip-flop also has three control inputs. The first control signal is the clock; it is used to synchronize the circuit. Other control signals are set and reset; these are used to set or reset the value of the corresponding flip-flops. These are also used to design registers, counters and other sequential circuits. There are four types of flip-flop.

2.1.1 S-R F/F

S-R is the basic F/F; it means other F/Fs can be designed using this F/F. It has two inputs and two outputs. It has an indeterminant condition (or undefined) when S = 1, R = 1. F/Fs are designed with the help of logic gates and the clock.

Digital VLSI Design and Simulation with Verilog, First Edition. Suman Lata Tripathi, Sobhit Saxena, Sanjeet Kumar Sinha, and Govind Singh Patel.
© 2022 John Wiley & Sons Ltd. Published 2022 by John Wiley & Sons Ltd.

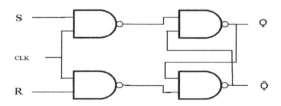

Figure 2.1　Clocked S-R F/F.

Table 2.1　Truth table of an S-R F/F.

I/P			O/P
Clk	S	R	Qt+
↑	0	0	Qt
↑	0	1	0
↑	1	0	1
↑	1	1	?

The S-R F/F designed with NAND-gate and with clock input is depicted in Figure 2.1.

A truth table of S-R F/F is given in Table 2.1.

The characteristic equation of S-R F/F is:

$$Q(t+) = SQt' + R'Qt \tag{1}$$

2.1.2　D F/F

D-F/F is similar to S-R with S and R input connected through a NOT gate. It is also called the delay F/F. It works as a buffer and produces output as input. It is generally used to design various types of registers. A block diagram of D-F/F is shown in Figure 2.2.

A truth table of D-F/F is given in Table 2.2.

The characteristic equation is:

$$Q(t+) = D \tag{2}$$

2.1.3　J-K F/F

J-K F/F also has two inputs. It is widely used to design counters and shift registers. There is no such undefined condition as in S-R F/F. For $J = 1$ & $K = 1$, it

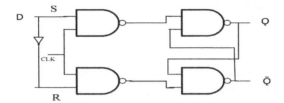

Figure 2.2 Clocked D-F/F.

Table 2.2 Truth table of a D-F/F.

Input		Output
Clk	D	Qt +
↑	0	0
↑	1	1

shows a toggle condition where the output will be a complimented form of the previous state that overcomes the limitation of S-R F/F. In this F/F, the inverted output is connected to its corresponding inputs. Figure 2.3 describes a gate level representation of J-K F/F.

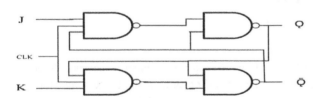

Figure 2.3 Clocked J-K F/F.

Table 2.3 Truth table of a J-K F/F.

Inputs			Output
Clk	J	K	Qn+1
↑	0	0	Qn
↑	0	1	0
↑	1	0	1
↑	1	1	Qn'

A truth table of J-K F/F is given in Table 2.3.

The characteristic equation of J-K F/F is:

$$Q(t+1) = JQ'(t) + K'Q(t) \qquad (3)$$

J-K F/F suffers with the race around condition due to the frequent toggle of output over a long duration clock compared to the propagation delay associated with a F/F. In this situation, F/F toggles more than once in the same clock cycle period. To prevent the race around condition in J-K F/F:

- The clock period should be kept lower than the F/F delay.
- The use of a Master F/F also nullifies the effect of such frequent toggles.

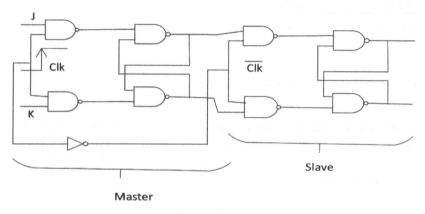

Figure 2.4 Master-slave of a J-K F/F.

The Master-slave J-K F/F shown in Figure 2.4 has two series connected S-R F/Fs where one acts as master and other as a slave. Here, master and slave circuits are triggered with a leading and falling edge of the clock respectively. This means master and slave F/Fs are active in different halves of the clock cycle.

2.1.4 T-F/F

This is referred to as a toggle (T) F/F. It has a single input. If input is zero, there is no change in output. But if input is 1, it toggles the previous state. This is generally used to design counters. A block diagram of T-F/F is shown in Figure 2.5.

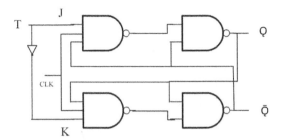

Figure 2.5 Clocked T-F/F.

Table 2.4 Truth table of a T-F/F.

I/O		O/P
Clk	T	Qn+1
↑	0	Qn
↑	1	Qn'

A truth table of T-F/F is given in Table 2.4.

The characteristic equation of T-F/F is:

$$Q(t+1) = TQ'(t) + T'Q(t)$$

2.1.5 F/F Excitation Table

An F/F excitation table gives information about the input values with the help of the current and next state of F/F output. Sometimes, an excitation table is preferred when only information about the output states is available. Figure 2.6 presents the different characteristics of F/F in a table.

2.1.6 F/F Characteristic Table

The F/F characteristic table gives information about the next state with the help of the current state and input values. Characteristic equations are derived from a characteristic table of F/F. Figure 2.7 shows the characteristic table of D- and T-F/F.

Qt	Qt+	S	R
0	0	0	X
0	1	1	0
1	0	0	1
1	1	X	0

a) S-R F/F

Qt	Qt+	T
0	0	0
0	1	1
1	0	1
1	1	0

c) T-F/F

Qt	Qt+	D
0	0	0
0	1	1
1	0	0
1	1	1

b) D-F/F

Qt	Qt+	J	K
0	0	0	X
0	1	1	X
1	0	X	1
1	1	X	0

d) T-F/F

Figure 2.6 Excitation table of a) S-R, b) D, c) J-K and d) D- F/F.

Qt	D	Qt+
0	0	0
0	1	1
1	0	0
1	1	1

a) D-F/F

Qt	T	Qt+
0	0	0
0	1	1
1	0	1
1	1	0

b) T-F/F

Figure 2.7 Characteristic table a) D-F/F and b) T-F/F.

2.2   Registers

A set of F/Fs is referred to as a register. A register with an n F/F unit can store 1-bit data or the information. In a few applications, these are also used to design counters. The clock is the most commonly used control signal used to control the process of the registers across various applications. The various types of the registers are described in next subsection.

2.2.1   Serial I/P and Serial O/P (SISO)

In SISO, digital information is shifted by one with every clock cycle and serially routed to the output node. The complete shifting of n bits is done in n number of clock cycles. It is used to store and shift data from right-to-left (left-shift) or left-to-right (right-shift) serially. This register takes four clock cycles to complete the 4-bit process. A SISO logic block is depicted in Figure 2.8. For example, if input data is 1010 then, after the second clock pulse, the register values will be 0010 for right-shift and 1000 for left-shift operations.

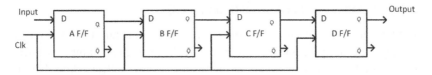

Figure 2.8   SISO shift register block diagram.

2.2.2   Serial Input and Parallel Output (SIPO)

Here, the data is shifted serially from one register to another and data can be taken from every F/F output parallelly. It also takes n clock cycles to complete the shift operation of an n-bit shift register. Figure 2.9 is a block diagram representation of a 4-bit SIPO using D-F/F.

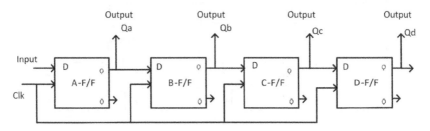

Figure 2.9   SIPO shift register block diagram.

2.2.3   Parallel Input and Parallel Output (PIPO)

PIPO is a parallel in and parallel output shift register where data can be applied or taken out parallelly. It is also known as a buffer register. It takes only one clock cycle to complete the process. Figure 2.10 is a block diagram representation of PIPO.

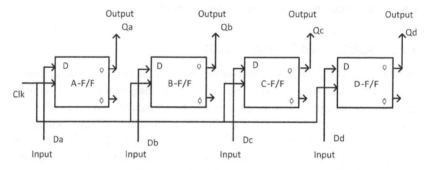

Figure 2.10   PIPO shift register block diagram.

2.2.4   Parallel Input and Serial Output (PISO)

In PISO, the I/P data is loaded to the registers in parallel but the O/P is taken serially. The design complexity increases with the greater number of logic blocks but data processing will be faster because of the parallel loading of input values. It has two blocks: the first block loads the input values to each register and the second enables the shifting of the data. The process of this register is controlled by load/shift mode input. Figure 2.11 shows a logic level representation of a 4-bit PISO using D-F/F and logic gates. We can replace logic gates with a 2×1 multiplexer block for the selection between load and shifting of data.

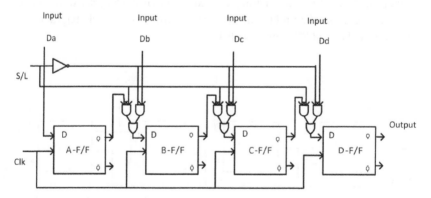

Figure 2.11   PISO shift register block diagram.

2.3 Counters

Counters are a sequential logic circuit used to count clock pulses arriving at clock I/P. They are also used in digital computer systems and other applications. There are two types of counters which are described in the next subsection.

2.3.1 Synchronous Counter

In a synchronous counter, the common clock is applied to all F/Fs simultaneously. Generally, T-or J-K F/Fs are used to design this type of counter. Because, if input of T-or J-K F/F is logic 1, then it toggles the previous state. A basic synchronous counter with three F/Fs have been presented in Figure 2.12.

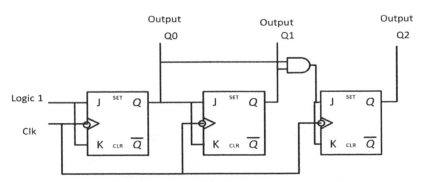

Figure 2.12 3-bit Synchronous up-counter with J-K F/F.

2.3.2 Asynchronous Counter

In this asynchronous, or ripple counter, an external clock is applied to the first F/F and the other F/Fs' clock input is connected to the O/P of the previous F/Fs. Generally, T-or J-K F/Fs are used to design asynchronous counters. Because, if the input of T-or J-K F/F is logic 1 then it toggles the previous state. The delay associated with a ripple counter will be more compared to a synchronous counter because of the different clock stages. Here, the next stage F/F has to wait for the clock input generated by previous block output. A 3-bit ripple counter is shown in Figure 2.13 where the inputs of each F/F are connected to logic 1.

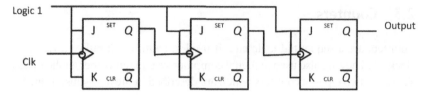

Figure 2.13   3-bit ripple counter (up-counter).

Table 2.5   State diagram of a 3-bit counter.

State	Q2	Q1	Q0
0	0	0	0
1	0	0	1
2	0	1	0
3	0	1	1
4	1	0	0
5	1	0	1
6	1	1	0
7	1	1	1

2.3.3   Design of a 3-Bit Synchronous Up-counter

In the synchronous counter, the clock applied to each F/F will be common and therefore there is less delay compared to a similar asynchronous counter. Table 2.5 and Figure 2.14 present the state diagram of a 3-bit synchronous up-counter that counts pulses in increasing order by one unit.

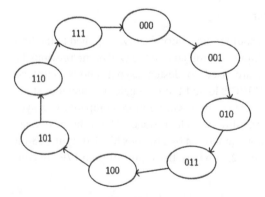

Figure 2.14   State diagram of a 3-bit up-counter.

An excitation table for the design is given in Table 2.6. Table 2.7 is used to design a 3-bit up-counter with the help of T-F/F.

Table 2.6 Excitation table of a T-F/F.

Qt	Qt+1	T
0	0	0
0	1	1
1	0	1
1	1	0

Table 2.7 State table of a 3-bit counter.

Output P S			N S			Input T-F/F		
Q2	Q1	Q0	Q2+	Q1+	Q0+	T2	T1	T0
0	0	0	0	0	1	0	0	1
0	0	1	0	1	0	0	1	1
0	1	0	0	1	1	0	0	1
0	1	1	1	0	0	1	1	1
1	0	0	1	0	1	0	0	1
1	0	1	1	1	0	0	1	1
1	1	0	1	1	1	0	0	1
1	1	1	0	0	0	1	1	1

K-map for T2:

Q2\Q1Q0	Q1'Q0'	Q1'Q0	Q1 Q0	Q1 Q0'
Q2'	0	0	1	0
Q2	0	0	1	0

Here, T2 can be expressed as: $T2 = Q1Q0$

K-map for T1:

Q2\Q1Q0	Q1' Q0'	Q1' Q0	Q1 Q0	Q1 Q0'
Q2'	0	1	1	0
Q2	0	1	1	0

Here, T1 can be expressed as: T1 = Q0

K-map for T0:

Q2\Q1Q0	Q1' Q0'	Q1' Q0	Q1 Q0	Q1 Q0'
Q2'	1	1	1	1
Q2	1	1	1	1

Here, for all minterms, the logic is 1 so, T0 = 1

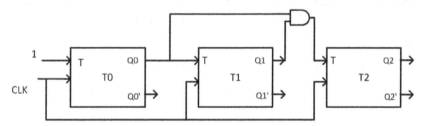

Figure 2.15   3-bit up-counter logic block.

Figure 2.15 shows a 3-bit up-counter using T-F/F. Similarly, we can design other types of counters using J-K F/F.

2.3.4   Ring Counter

A ring counter forms a loop with last stage F/F output connected to the input of the first stage F/F. A 4-bit ring counter can be designed with the help of four D-F/Fs. A ring counter is being used to determine various patterns or number values from a set of information. Figure 2.16 shows a 4-bit ring counter using D-F/F. Here, the counter must start with any non-zero value. Suppose the initial value given is 0001, then the sequence will be 1000, 0100, 0010, 0001, and will repeat the same sequence again.

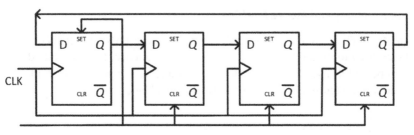

Figure 2.16   4-bit ring counter using D-F/F.

2.3.5   Johnson Counter

This is another type of counter which is being used to detect a particular pattern from given information. It consists of 4 days of F/Fs. It is a synchronous-type counter. Figure 2.17 shows a 4-bit Johnson counter where the complimentary out of the last stage F/F will be connected to the input of the first stage F/F. If the data loaded in F/F is 0000, then the sequence obtained will be 1000,1100,1110,1111, 0111, and so on with different clock edges.

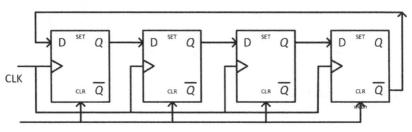

Figure 2.17   4-bit Johnson counter using D-F/F.

2.4   Finite State Machine (FSM)

FSM is a type of mathematical model. It is used to extract one number of states at any given time. It consists of the state diagram, state table, excitation table, and circuit diagram. This type of finite state machine is used in vending machines, elevators, and such like.

2.4.1 Mealy and Moore Machine

In Mealy FSM, the output depends upon external inputs and feedback. It is being used to detect patterns of a given input sequence. In this method, the number of bits used in the detection of the sequence is equal to the number of states used. Figure 2.18 presents the Mealy model block.

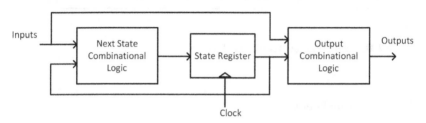

Figure 2.18 Mealy machine.

Mealy FSM is used to design a sequential circuit, for example, F/Fs, shift registers, counters, and detectors. Figure 2.19 depicts the Moore model.

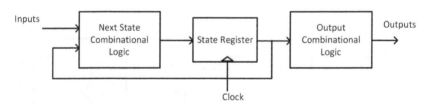

Figure 2.19 Moore machine.

2.4.2 Pattern or Sequence Detector

A sequence or pattern detector is a device which is used to detect sequence from input sequences. A clock is used to monitor sequence and display output in terms of 1 or 0. Figure 2.20 describes the block diagram of the binary sequence detector. There are two types of detector methods.

a) Overlapping
b) Non-overlapping

Let us take 011 for detection. If the number of bits are three, then states will be three in the case of a Mealy machine. It is a non-overlapping method. Figure 2.21 describes the state diagram of the Moore model. If the number of

Figure 2.20 Design 011 sequence using a Mealy machine.

bits in pattern or sequence are N, then the number of states required will be N in Mealy and N+1 in the Moore sequence detector.

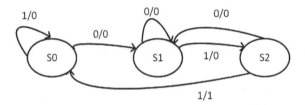

Figure 2.21 State-diagram representation of Moore model.

Here, S0, S1, S2 states are taken as: S0 = 00, S1 = 01, S2 = 11. Tables 2.8 and 2.9 show excitation and state table of D-F/F respectively.

Table 2.8 D-F/F excitation table.

Qt	Qt+1	D
0	0	0
0	1	1
1	0	0
1	1	1

Table 2.9 State table 1 of sequence 011.

PS		NS		O/P	
		X = 0	X = 1	X = 0	X = 1
Q1	Q2	Q1'	Q2'		
0	0	0	1	0	0
0	1	0	1	0	0
1	0	0	1	0	1

Table 2.10 State table 2 of sequence 011.

P S		State	N S		Flip-Flop		O/P
Q1	Q2	X	Q1'	Q2'	D1	D2	Y
0	0	0	0	1	0	1	0
0	0	1	0	0	0	0	0
0	1	0	0	1	0	1	0
0	1	1	1	0	1	0	0
1	0	0	0	1	0	1	0
1	0	1	0	0	0	0	1

Table 2.10 gives D-F/F input and output values depending on the present and next state of provided sequence.

K-map for D1:

Q1\Q2X	Q2' X'	Q2' X	Q2 X	Q2 X'
Q1'	0	0	1	0
Q1	0	0	X	X

D1 = Q2X (1)

K-map for D2:

Q1\Q2X	Q2' X'	Q2' X	Q2 X	Q2 X'
Q1'	1	0	0	1
Q1	1	0	X	x

D2 = X' (2)

K-map for Y:

Q1\Q2X	Q2' X'	Q2' X	Q2 X	Q2 X'
Q1'	0	0	0	0
Q1	0	1	X	x

Y = Q1X (3)

The above three equations are used to design the sequence circuit. Where X is I/P and Y is O/P.

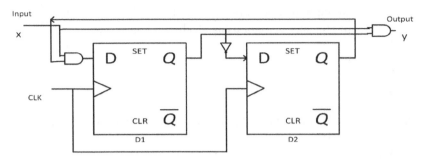

Figure 2.22   Sequence circuit of 011 Mealy sequences.

The circuit diagram of a given 011 sequence is shown in Figure 2.22.

Review Questions

Q1 Draw the truth table and Excitation of D and T flip-flops. Compare them with SR and JK respectively.

Q2 Design and implement a 3-bit synchronous down-counter using flip-flop.

Q3 Draw the truth table and logic diagram of D and T flip-flop. Give one application of each flip-flop.

Q4 Convert a J-K flip-flop into a D flip-flop.

Q5 Design and implement a 3-bit asynchronous up-counter using flip-flop.

Q6 Draw an excitation table and logic diagram of SR and JK flip-flops using a NOR gate.

Multiple Choice Questions

Q1 The minimum number of states required in a 3-bit input sequence detector using Mealy and Moore models are:
  A 3,3
  B 3,4
  C 4,3
  D 4,4

Q2 The minimum number of flip-flops required in the implementation of a 3-bit input sequence detector using Mealy and Moore models are:

A 2,2

B 2,3

C 3,2

D 3,3

Q3 The minimum number of T flip-flops required to implement a mod10 synchronous counter are:

A 3

B 4

C 5

D None of the above

Q4 How many natural states will there be in a 4-bit ripple counter?

A 4

B 8

C 16

D 32

Q5 Which statement is correct for a synchronous counter compared to an asynchronous counter?

A Fast, less hardware, same clock to all flip-flops

B Slow, less hardware, different clock to all flip-flops

C Fast, less hardware, different clock to all flip-flops

D Slow, more hardware, same clock to all flip-flops

Q6 The output of a 4-bit SISO shift register after three clock pulses for the input sequence 110011 is:

A 1100

B 0011

C 0110

D 1001

Q7 The output of the 4-bit PIPO shift register after first clock pulse for the input sequence 1001 is:

A 1000

B 0001

C 0101

D 1001

Reference

[1] Morris, M.M., and Ciletti, M.D. (2018). Digital Design. Upper Saddle River, NJ: Pearson.

3

Introduction to Verilog HDL

3.1 Basics of Verilog HDL

Verilog is a type of hardware description language (HDL). This language is used to describe the hardware for the purpose of simulation, synthesis, and implementation. Verilog describes a circuit as an N/W switch, ROM, RAM, micro-controller, micro-processors, and other combinational and sequential circuits [1–2]. HDLs can be described at any level of abstraction.

3.1.1 Introduction to VLSI

Very large-scale integration (VLSI) is a type of integration. Millions of transistors are fabricated in one single IC under this category. Microprocessors and micro-controllers also come under this VLSI category. Due to the complexity of the circuitry, these types of circuits are not possible to verify, fabricate, or breadboard. These circuits require a special type of tool or software for verification and fabrication. It is very difficult to design computer-aided techniques (CAD) with VLSI digital circuits. Therefore, VLSI-CAD has been introduced to solve these types of complexity.

3.1.2 Analog and Digital VLSI

VLSI is also an HDL. It is used to simulate digital and analog signals. But here, we will discuss only digital design. Analysis and design of analog mixed signal (AMS) is not included in this book. The basics of design and analysis of HDLs are described in this chapter.

Digital VLSI Design and Simulation with Verilog, First Edition. Suman Lata Tripathi, Sobhit Saxena, Sanjeet Kumar Sinha, and Govind Singh Patel.
© 2022 John Wiley & Sons Ltd. Published 2022 by John Wiley & Sons Ltd.

3.1.3 Machine Language and HDLs

Machine language and HDLs are those languages used in digital design to perform all types of operations. We know that all operations are performed by machine language in a digital system. These two languages play a very important role in the design of digital systems. These languages are also used to design high-frequency synthesizers for communications. The next subsection describes the design methodologies of any circuit for simulation and synthesis.

3.1.4 Design Methodologies

Design methodology is an initial stage which represents design methodology sub-blocks. There are two types of design methodologies.

i) Top-down design methodology
In this design methodology, we can define top-level design block and then sub-level blocks. Further, sub-blocks are divided into leaf-cells. These leaf-cells cannot be further divided. This design method is called top-down design methodology (Figure 3.1).

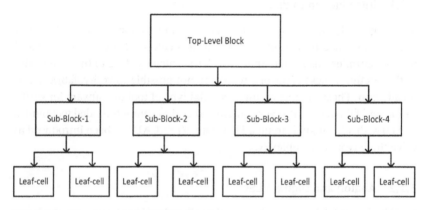

Figure 3.1 Top-down design methodology.

ii) Bottom-up design methodology
In this design methodology, we can define the bottom-level design block available to us. With the help of leaf-cells, we can build bigger macro cells. These bigger cells can be used to design higher levels of blocks. This design method is called bottom-up design methodology (Figure 3.2).

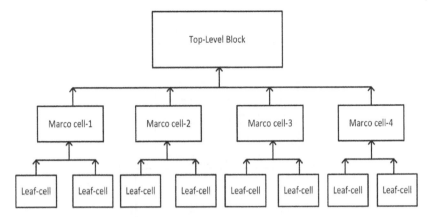

Figure 3.2 Bottom-up design methodology.

3.1.5 Design Flow

This section describes the design flow of the VLSI IC circuits (Figure 3.3). The design specification is the main parameter to design any VLSI IC circuit. First, the rectangular block describes the specifications of the design. Next, the block describes the behavior of the design to be used in HDL. A functional verification block is used to verify RTL synthesis and simulation. It also tests Verilog code for logic synthesis. The next block is the preparation of a gate level netlist for logical verification and testing. If testing or verification fails, then codes will be tested again. Floor planning, automatic place and route are important procedures before the physical layout. The next step is physical layout; once this is complete, the next step is layout verification. And, finally, the last and final step to complete the process is implementation. If the layout is verified, we can say that the circuit can be fabricated on chip.

3.2 Level of Abstractions and Modeling Concepts

Verilog is an HDL. Its behavior is similar to behavioral and structural language. Its internal module can be divided into four levels of abstraction. The details of each module are described in the next subsections.

3.2.1 Gate Level

In the gate level of abstraction, a module is designed with logic gates and these gates are interconnected through nets. A description of this level is

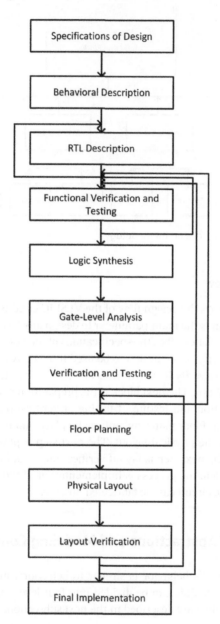

Figure 3.3 Design flow chart.

similar to the gate-level diagram. When circuits are simple, this level of abstraction is used for design.

3.2.2 Dataflow Level

In the dataflow level of abstraction, specific dataflow is used for design. In this method, we know how dataflows between hardware registers and how data will be processed in the circuit.

3.2.3 Behavioral Level

A behavioral level of abstraction is also called an algorithmic level. The behavioral level is the highest level of abstraction. Using this method, a module can be designed using the desired algorithm without concern about hardware details. This level of abstraction is similar to programing in C language.

3.2.4 Switch Level

Switch level is the lowest level of abstraction. In this, design can be implemented among the switches, storage nodes, interconnecting wires, and input/output blocks. With this method, we should have knowledge about switch-level implementation in Verilog. It is difficult to design complex circuits using this method. This method is generally used to design low-power devices.

3.3 Basics (Lexical) Conventions

This language has a stream of lexical token which can be used as: numbers, identifiers, keywords, comments, or strings, etc. These tokens, used in Verilog, are similar to programing in C. It is case-sensitive language and, in this language, all keywords are in lowercase. A few conventions are described in the next subsection.

3.3.1 Comments

Comments are used to understand the program and documentation. There are two types of comments in Verilog. ZFirst is a one-line comment and the next are multiple-line comments.

Y = A & B;   //This is an example of a one-line comment
Z = #10(B | C);   //time parameter
/* this is a multiple
line comment */
/* This is the testbench program
module tb();
reg a,b;
output y;
*/

3.3.2   Whitespace

There are three types of whitespace used in Verilog.

a)   Blank spaces (\b)
b)   Tabs (\t)
c)   Newlines (\n)

In Verilog, whitespace is not ignored in strings.

3.3.3   Identifiers

Identifiers are used as reference names given to design so that objects can be used later in the program. These are made up from underscore (_), dollar sign ($), alphanumeric characters, and some special symbols. Identifiers cannot be started with special symbol like $, _ or numbers.

Example:

input clock;   //clock in identifier

output q;   //q is identifier

3.3.4   Escaped Identifiers

These are an expanded type of identifier. They begin with a backslash(\) special character and end with any whitespace. Whitespace and backspace cannot be considered as identifiers.

Example:

\w-x+y+z

*LPU_Jalandhar

3.3.5   Keywords

Keywords are case sensitive. These must be written in lowercase.

Example:

input a, b;	//input is keywords
reg out;	//reg is keywords
INPUT clock;	//Illegal keywords, it must be in lowercase

3.3.6 Strings

These are a sequence of characters. String can be declared in double quotes only. However, it must be on a single line.

Example:

"Welcome to LPU Jalandhar"//it is a string

"xyz/abc"//it is string

3.3.7 Operators

There are three types of operator. These operators are used to perform binary operations.

Unary operation:

It precedes the operand. The symbol is (~).

Example:

Y = ~X;

Z = ~X && W;

Binary operation:

Binary operations perform between two operands.

Example:

Z = X && Y;

C = A | B;

Ternary operation:

Ternary operators have three operands.

Example:

Z = W ? X: Y;

3.3.8 Numbers

Numbers are used to perform various operations. There are two types of numerical systems.

Sized numbers

There is a format for sized numbers. This is used to perform these operations.

Syntax for a sized number:

<size>' <base format> <number>

Example:

D or d = Decimal numbers = 0, 1, 2......9

O or o = Octal numbers = 0, 1.........7

B or b = Binary numbers = 0,1

H or h = Hexadecimal numbers = 0, 1, 2......9, a,....e, f

Example:

4′b1010;	//4-bit binary number
8′hbc;	//8-bit hexadecimal number
16′d20;	//16-bit decimal number

Unsized numbers

Unsized numbers are those numbers written without any format. They have a 32-bit default value.

Example:

12345;	//32-bit decimal number by default
'ha2;	//32-bit hexadecimal number
'o23;	//32-bit octal number

3.4   Data Types

Data types are used to perform various operations. There are many types of data type.

3.4.1   Values

Verilog support only four values.

0 -> Logic zero, false condition

1 -> Logic one, true condition

x -> Unknown value

z -> High impedance, floating state

3.4.2   Nets

Nets connect the hardware elements. They are also called wires. Wire is a keyword. It connects inputs and output with its internal circuitry.

Example:

wire x;	//declare as a net or wire
wire = 1′b1;	//wire has 1 fixed value

3.4.3 Registers

Registers are an element of data storage, which store data temporarily for the purpose of simulation. These are hardware registers built from triggers in real circuits. They are keyword registered.

Example:

reg clear;   //variable clear can hold its value

initial

begin

clear = 1'b0;   //initialize clear to 0

#10 clear = 1'b1;

end

3.4.4 Vectors

Multiple bit widths can be represented as vectors. Reg or nets can also be represented as vectors.

These can be declared in the following format; [high#: low] or [low#: high#]. The left number in brackets will be the most significant bit (MSB) of the vector.

Example:

wire x;   //it is scalar net

wire [7:0] y;   //it is 8-bit bus

reg [0:31] addr;   //it is 32-bit bus

3.4.5 Integer Data Type

This is represented as the keyword **integer**. In this data type, we can only use integer value.

Example:

integer count;   //it is as integer data type

initial

   count = 1;   //

3.4.6 Real Data Type

This is represented as the keyword **real**. In this data type, we can only use real value.

Example:

real del;   //it is a real data type

initial

```
begin
    del = 4e8;        //real no. notation
    del = 2.13;       //del assigned a value 2.13
end
integer i;
integer
    i = del;          //it is integer value so it get the value 2(rounded value)
```

3.4.7 Time Data Type

This is represented as the keyword **time**. In this data type, we can only use time parameters.

Example:

```
time delay;           //define time
initial
    delay = $time;    //it saves the current simulation time
```

3.4.8 Arrays

The use of Array in Verilog is similar to C programing. It is allowed for integer, time, reg, and vectors. We cannot take array for real numbers.

Example:

```
integer count[7:0];
time point[1:100];
reg bus[31:0];
```

3.4.9 Memories

Memories can be represented as an Array of registers in Verilog. Each element of an Array is also called a Word. A Word may be one or more bits. In simulation, RAMs, ROMs, and other register files can be designed.

```
reg mes_size[0:1023];
reg [7:0] mem_addr[0:1023];
```

3.5 Testbench Concept

The testbench concept is also called a stimulus. The main module is used to generate RTL of the design. But testbench is also used to generate wave to verify output of the design. The main module may be changed as per each level of abstraction, whereas test is the same for all types of modeling. We can verify our output using testbench. Also, we can analyze the timing parameters of the design.

Multiple Choice Questions

Q1 The Verilog keyword of system task that displays every change in output for the input changes is:
 A $display
 B $monitor
 C $finish
 D $stop

Q2 The incorrect representation of a sized number in Verilog is:
 A 4'b1111
 B 12'habc
 C 'hc3
 D 16'd25x

Q3 The correct syntax in Verilog for a system task of display is:
 A $display ("Hello Verilog World");
 B $display('Hello Verilog World');
 C $display "Hello Verilog World";
 D $display(time "Hello Verilog World");

Q4 The tasks of frontend and backend design engineers are:
 A Functional Design & Physical Verification respectively.
 B Physical Verification & Functional Design respectively.
 C Functional Verification & Static Timing Analysis respectively.
 D Both 1 and 3.

Q5 Which is not true in Veriog HDL?
  A Verilog HDL allows different levels of abstraction to be mixed in the same models.
  B Verilog is case sensitive.
  C Verilog is based on Pascal and Ada.
  D Designer can define the hardware model in terms of switches, gates, RTL, or behavioral code.

Q6 Which statement is true for test stimulus in Verilog HDL?
  A The stimulus block instantiates the design block and directly drives the signals in the design block.
  B Stimulus is to instantiate both the stimulus and design blocks in a top-level dummy module.
  C The stimulus block interacts with the design block only through the interface.
  D All of the above.

Q7 The inclusion of a *logic.v* file in to another *design.v* file can be done by using keyword:
  A 'define logic.v
  B 'include logic.v
  C 'define S logic.v
  D 'include S logic.v

References

[1] Palnitkar, S. (2001). Verilog HDL. Upper Saddle River, NJ: Pearson.
[2] Warkley, J.F. (2005). Digital Design: Principles and Practices. Upper Saddle River, NJ: Pearson.

4

Programming Techniques in Verilog I

4.1 Programming Techniques in Verilog I

Verilog HDL is a synthesized tool through which circuit designers can design their desired circuit in their own way. In Verilog, the digital circuit can be described in terms of a network of digital components. Verilog programming has the same C language type syntax. Verilog is used to describe hardware whereas C language is considered software language. In Verilog, the statements are concurrent compared to other languages. In other languages, the statements execute sequentially. The basic building block for writing Verilog code is the module statement. In a module definition, all input as well as output of the desired circuit is defined. Synthesized Verilog codes are mapped to the actual hardware logic gates of the circuits. In this chapter, the gate-level modeling of the circuit is described with the help of different types of circuit.

4.2 Gate-Level Model of Circuits

In gate-level modeling, logic gates used in the circuits are called from the available library of Verilog HDL. In this modeling, the logic gates used have one-to-one relations inside the hardware schematic. The gate-level model is considered as the lowest level of design model.

Digital VLSI Design and Simulation with Verilog, First Edition. Suman Lata Tripathi, Sobhit Saxena, Sanjeet Kumar Sinha, and Govind Singh Patel.
© 2022 John Wiley & Sons Ltd. Published 2022 by John Wiley & Sons Ltd.

Example 1: A basic logic circuit at gate level is shown in Figure 4.1.

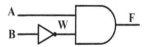

Figure 4.1 Logic circuit.

Verilog Code:

module Circuit(A,B,F);

                                   input A, B;
                                   output F;
                                   wire W;
                                   not (W,B);
                                   and (F,A,W);

endmodule

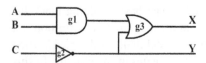

Figure 4.2 Logic circuit.

Example 2: A logic circuit with AND, OR, and NOT gate is shown in Figure 4.2.

Verilog Code:

module circuit(A,B,C,X,Y);

                                   input A,B,C;
                                   output X,Y;

wire w;

                                   and g1(w,A,B);
                                   not g2(Y, C);
                                   or g3(X,w,Y);

endmodule

4.3   Combinational Circuits

In digital logic design, the circuit where output depends on the combination of present inputs, and does not depend upon previous inputs, is called a combinational circuit. It consists of logic gates, input variables, and output variables.

Gate-level modeling of a combinational circuit

- Adder and Subtractor
- Multiplexer and De-multiplexer
- Decoder and Encoder
- Comparator

4.3.1   Adder and Subtractor

4.3.1.1   Adder
In adder, irrespective of inputs, there are always two outputs in terms of sum and carry. A half adder has two inputs whereas in the case of a full adder, the number of inputs becomes three.

Half adder
A and B are the two inputs defined in the half adder whereas S and C are considered outputs. Figure 4.3 shows the block diagram of a half adder circuit. As there are only two inputs, input combinations are four e.g., 00, 01, 10, 11. All possible four combinations, as well as corresponding outputs, are described in Table 4.1.

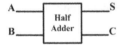

Figure 4.3   Block diagram of a half adder.

Table 4.1   Half adder.

Inputs		Outputs	
A	B	S	C
0	0	0	0
0	1	1	0
1	0	1	0
1	1	0	1

Based on the observations given in Table 4.1, one 2-input XOR gates, along with one 2-input AND gate, are required for implementation of the half adder circuit. Figure 4.4 shows the logic diagram of a half adder circuit. The Boolean expression for a half adder is defined as follows:

$$S = A'B + AB' = A \oplus B \text{ and } C = AB$$

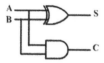

Figure 4.4 Logic circuit of half adder.

Verilog Code:
In Verilog code of the half adder using gate-level modeling, inputs and outputs are defined inside the module HA. Inside the module, two logic gates are called.

```
module HA(A, B, S, C);
    input A,B;
    output S,C;
    xor X_2(S,A,B);
    and A_2(C, A,B);
endmodule
```

Half Adder using a NAND gate:
A minimum of five NAND gates are required for a half-adder circuit. The circuit using five NAND gates is described in Figure 4.5.

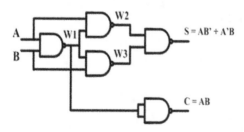

Figure 4.5 Logic circuit of half adder using a NAND gate.

Verilog Code:
In Verilog code of a half adder using NAND, inputs and outputs are defined inside the module half adder. Inside the module, five NAND gates are called along with three wires.

```
module HA(A, B, S, C);
    input A,B;
    output S,C;
    wire W1,W2,W3;
    nand N_1(W1,A,B);
    nand N_2(W2,A,W1);
    nand N_3(W3,W1,B);
    nand N_4(S,W2,W3);
    nand N_5(C,W1,W1);
    endmodule
```

Half Adder using NOR gate:

Implementation of the half adder using NOR gates only requires five minimum gates. The circuit using five NOR gates is shown in Figure 4.6.

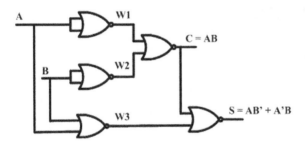

Figure 4.6 Logic circuit of half adder using NOR gate.

Verilog Code:

In Verilog code of half adder using NOR, inputs and outputs are defined inside the module half adder. Inside the module, five NOR gates are called along with three wires.

```
module HA(A, B, S, C);
    input A,B;
    output S,C;
    wire W1,W2,W3;
    nor N_1(W1,A,A);
    nor N_2(W2,B,B);
    nor N_3(W3,A,B);
    nor N_4(C,W1,W2);
    nor N_5(S,W3,C);
    endmodule
```

Full Adder

In a full-adder circuit, A, B, and C are considered as three inputs whereas S and C are considered as output. As there are only three inputs, the input combinations are eight: 000, 001, 010, 011, 100, 101, 110, 111. All possible four combinations, as well as corresponding outputs, are described in Table 4.2. Figure 4.7 presents a block diagram of a full adder.

Figure 4.7 Block diagram of a full adder.

Table 4.2 Full adder.

Inputs			Outputs	
A	B	Cin	S	Co
0	0	0	0	0
0	0	1	1	0
0	1	0	1	0
0	1	1	0	1
1	0	0	1	0
1	0	1	0	1
1	1	0	0	1
1	1	1	1	1

Based on the observations given in Table 4.2, one 3-input XOR gate, along with three 2-input AND gates, and one 3-input OR gates, are required for implementation of the full-adder circuit. Figure 4.8 presents the logic diagram of a full adder. The expression in terms of three input variables of full adder is defined as follows [1].

$S = A'B'Cin + A'BCin' + AB'Cin' + ABCin = A \oplus B \oplus Cin$
and $Co = A'BCin + AB'Cin + ABCin' + ABCin = AB + BCin + ACin$

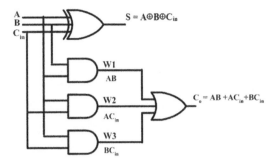

Figure 4.8 Logic circuit of a full adder.

Verilog Code:
In Verilog code of the full adder using gate-level modeling, inputs and outputs are defined inside the module FA. Inside the module, one XOR gate, three AND gates, as well as one OR gates are used.

```
module FA(A, B, Cin, S, Co);
    inputA,B,Cin;
    outputS,Co;
    wire W1,W2,W3;
    xor X_1(S,A,B,Cin);
    and A_1(W1,A,B);
    and A_2(W2,A,Cin);
    and A_3(W3,B,Cin);
    or O_1(Co,W1,W2,W3);
endmodule
```

4.3.1.2 Subtractor
In subtractor, irrespective of the inputs, there are always two outputs in terms of difference and borrow. In the case of the half subtractor, the number of inputs is two whereas in the case of the full subtractor, the number of inputs becomes three.

Half Subtractor
In the half-subtractor circuit, A and B are two inputs whereas D and B_0 are considered as output. As there are only two inputs, the input combinations are four: 00, 01, 10, 11. All possible four combinations, as well as corresponding outputs, are described in Table 4.3. Figure 4.9 presents a half-subtractor block diagram.

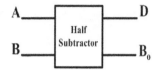

Figure 4.9 Block diagram of a half subtractor.

Table 4.3 Half subtractor.

Inputs		Outputs	
A	B	D	B_0
0	0	0	0
0	1	1	1
1	0	1	0
1	1	0	0

Based on the observations given in Table 4.3, one 2-input XOR gate, along with one 2-input AND gate and one NOT gate, are required for implementation of a half-subtractor circuit. Figure 4.10 presents the logic diagram of a half subtractor. The Boolean expression for the half subtractor is defined as follows [1].

$$D = A'B + AB' = A \oplus B \text{ and } Bo = A'B$$

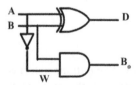

Figure 4.10 Logic circuit of a full subtractor.

Verilog Code:
In Verilog code of the half subtractor using gate-level modeling, inputs and outputs are defined inside the module HS. Inside the module, three logic gates as well as one wire is called.

```
module HS(A, B, D, Bo);
input A,B;
```

outputD,Bo;

wire W;

xor X_2(D,A,B);

not N_1(W,A)

and A_2(Bo, W,B);

endmodule

Half Subtractor using NAND gate:

Implementation of the half subtractor using NAND gates only requires a minimum of five NAND gates. The circuit using five NAND gates is described in Figure 4.11.

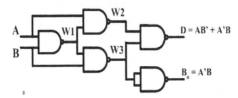

Figure 4.11 Logic circuit of a half subtractor using a NAND gate.

Verilog Code:

In Verilog code of the half subtractor using NAND, inputs and outputs are defined inside the module HS. Inside the module, five NAND gates are called along with three wires.

module HS(A, B, D, Bo);

input A,B;

output D,B0;

wire W1,W2,W3;

nand N_1(W1,A,B);

nand N_2(W2,A,W1);

nand N_3(W3,W1,B);

nand N_4(D,W2,W3);

nand N_5(Bo,W3,W3);

endmodule

Half Subtractor using NOR gate:

Implementation of the half subtractor using only NOR gates requires a minimum of five gates. The circuit using five NOR gates is described in Figure 4.12.

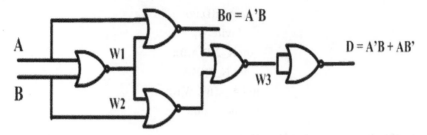

Figure 4.12 Logic circuit of the half subtractor using a NOR gate.

Verilog Code:
In Verilog code of a half subtractor using NOR, inputs and outputs are defined inside the module HS. Inside the module, five NOR gates are called along with three wires [2].

```
module HS(A, B, D, Bo);
    input A,B;
    outputD,Bo;
    wire W1,W2,W3;
    nor N_1(W1,A,B);
    nor N_2(W2,W1,B);
    nor N_3(Bo,A,W1);
    nor N_4(W3,Bo,W2);
    nor N_5(D,W3,W3);
endmodule
```

Full Subtractor
In a full-subtractor circuit A, B, and C are considered as three inputs whereas D and Bo are assumed to be output. As there are only three inputs, the input combinations are eight: 000, 001, 010, 011, 100, 101, 110, 111. All possible four combinations, as well as corresponding outputs, are described in Table 4.4. Figure 4.13 shows the full-subtractor block diagram.

Figure 4.13 Block diagram of a full subtractor.

Table 4.4 Full subtractor.

Inputs			Outputs	
A	B	C	D	B_o
0	0	0	0	0
0	0	1	1	1
0	1	0	1	1
0	1	1	0	1
1	0	0	1	0
1	0	1	0	0
1	1	0	0	0
1	1	1	1	1

Based on the observations given in Table 4.4, one 3-input XOR gate, along with three 2-input AND gates, as well as one 3-input OR gate and one NOT gate are required for implementation of a full-subtractor circuit. Figure 4.14 presents the logic diagram of a full subtractor. The Boolean expression for the full subtractor is defined as follows.

$$D = A'B'Cin + A'BCin' + AB'Cin' + ABCin = A \oplus B \oplus Cin$$
$$\text{and } Bo = A'B'Cin + A'BCin' + A'BCin + ABCin = A'B + BCin + ACin$$

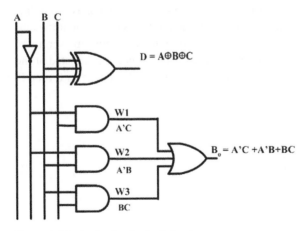

Figure 4.14 Logic circuit of a full subtractor.

Verilog Code:

In Verilog code of the full subtractor using gate-level modeling, inputs and outputs are defined inside the module FS. Inside the module, one XOR gate, three AND gates as well as one OR gate are used [2].

```
module FS(A, B, C, D, Bo);
    input A,B,C;
    outputD,Bo;
    wire W1,W2,W3,W4;
    xor X_1(D,A,B,C);
    not n_1(W1,A);
    and A_1(W2,W1,C);
    and A_2(W3,W1,B);
    and A_3(W4,B,C);
    or O_1(Bo,W2,W3,W4);
endmodule
```

4.3.2 Multiplexer and De-multiplexer

4.3.2.1 Multiplexer

The multiplexer is another combinational circuit which is an important part of digital electronics. Multiplexers are known as data-selector devices, which means they only select one output at a time from the many available inputs based on the select lines [1].

2 × 1 Multiplexer

In a 2 × 1 multiplexer, there are two inputs; one select line and one output line. Based on the select line, the inputs are reflected in terms of output at the output terminal. As only one select line is present in the 2 × 1 multiplexer, there are two options—0 and 1. As shown in Figure 4.15, when 0 is selected at the select line, the I0 is reflected at Y, and when the select line becomes 1, then I1 is considered as output [1].

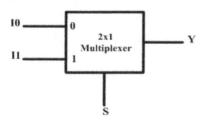

Figure 4.15 Block diagram of a 2 × 1 multiplexer.

Table 4.5 2 × 1 multiplexer.

Select line	Output
S	Y
0	I0
1	I1

Based on the observations given in Table 4.5, the Boolean expression for a 2 × 1 multiplexer is defined as follows:

Y = S' I0 + SI1

Circuit diagram

In the implementation of a 2 × 1 multiplexer, one NOT gate along with two 2-input AND gates, as well as one 2-input OR gates are required [3]. Figure 4.16 presents the logic diagram of a 2 × 1 multiplexer.

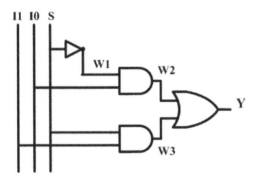

Figure 4.16 Logic circuit of a 2 × 1 multiplexer.

Verilog Code

In Verilog code of a 2 × 1 multiplexer using gate-level modeling, inputs and outputs are defined inside the module mux_2x1. Inside the module, one NOT gate, two AND gates as well as one OR gate, along with three wires are used [2].

```
module mux_2x1(Y,I0,I1,S);
    input I0,I1,S;
    output Y;
    wire W1,W2,W3;
```

not (W1,S);

and (W2,I0,W1);

and (W3,I1,S);

or (Y,W2,W3);

endmodule

4 × 1 Multiplexer

In the 4 × 1 multiplexer, there are four inputs, two select lines, and one output line. Based on the select lines, the inputs are reflected in terms of output at the output terminal. As only two select lines are present in a 4 × 1 multiplexer, there are four options: 00, 01, 10, 11. As shown in Figure 4.17, when the select line becomes 00, the I0 is reflected at output port Y, and when the select line becomes 11, then I3 is considered as output [3].

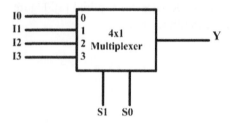

Figure 4.17   Block diagram of a 4 × 1 multiplexer.

Table 4.6   4 × 1 multiplexer.

Select lines		Output
S1	S0	Y
0	0	I0
0	1	I1
1	0	I2
1	1	I3

Based on the observations given in Table 4.6, the Boolean expression for the 4 × 1 multiplexer is defined as follows:

$$Y = S1'S0'I0 + S1'S0I1 + S1S0'I2 + S1S0I3$$

Circuit diagram

Two NOT gates, four 3-input AND gates along with one 4-input OR gate are required for the 4 × 1 multiplexer circuit [3] as shown in Figure 4.18.

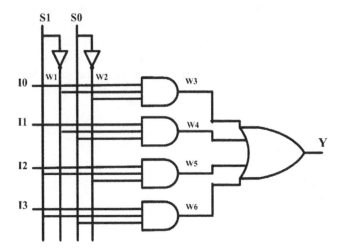

Figure 4.18 Logic circuit of a 4 × 1 multiplexer.

Verilog Code:

In Verilog code of a 4 × 1 multiplexer using gate-level modeling, inputs and outputs are defined inside the module mux_4x1. Inside the module, two NOT gates, four AND gates, as well as one OR gate, along with six wires are used.

```
module mux_4x1(Y,I0,I1,I2,I3,S1,S0);

    input I0,I1,I2,I3,S1,S0;
    output Y;
    wire w1,w2,w3,w4,w5,w6;
    not (w1,S1);
    not (w2,S0);
    and (w3,I0,w1,w2);
    and (w4,I1,w1,S0);
    and (w5,I2,w2,S1);
    and (w6,I3,S0,S1);
    or (Y,w3,w4,w5,w6);
    endmodule
```

4.3.2.2 De-multiplexer

The de-multiplexer is a combinational circuit known as a data distributor. In this circuit, a single input is selected through the select line into multiple outputs.

1 × 2 De-multiplexer

In a 1 × 2 de-multiplexer, there is one input, one select line and two output lines. Based on the select line, inputs are reflected in terms of output at the output terminal. As only one select line is present in the 1 × 2 de-multiplexer, there are two options; 0 and 1. As shown in Figure 4.19, when 0 is selected at the select line, D is reflected at output port Y0, and when the select line becomes 1, then input D is reflected at port Y1.

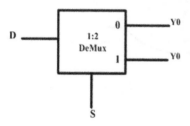

Figure 4.19 Block diagram of 1 × 2 de-multiplexer.

Table 4.7 1 × 2 de-multiplexer.

Select line	Input	Outputs	
S	D	Y1	Y0
0	0	0	0
0	1	0	1
1	0	0	0
1	1	1	0

Based on the observations given in Table 4.7, the Boolean expression for a 1 × 2 de-multiplexer is defined as follows:

$$Y0 = DS'$$
$$Y1 = DS$$

Circuit diagram

In the implementation of a 1 × 2 de-multiplexer, one NOT gate along with two 2-input AND gates are required. A 2 × 2 de-multiplexer logic diagram is shown in Figure 4.20.

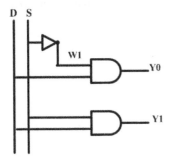

Figure 4.20 Logic circuit of a 1 × 2 de-multiplexer.

Verilog Code:

In Verilog code of 1 × 2 de-multiplexer using gate-level modeling, inputs and outputs are defined inside the module Dmux_1x2. Inside the module, one NOT gate, as well as two AND gates along with one wire is used.

```
module Dmux_1x2(D,S,Y0,Y1);
    input D,S;
    output Y0,Y1;
    wire w1;
    not (w1,s);
    and (Y0,w1,D);
    and (Y1,S,D);
endmodule
```

4.3.3 Decoder and Encoder

4.3.3.1 Decoder

The decoder circuit is a combinational circuit having n input gate with 2N number of output gates. If the number of inputs becomes n and the number of output is m then:

m = 2N

2-to-4 Decoder

In the 2-to-4 decoder, there are two input lines as well as four output lines. Based on the selection of input lines, outputs are displayed at the output terminal. As two input lines are present, there are a total of four possibilities: 00, 01, 10, 11. Figure 4.21 shows that when the input lines become 00, the output pin D0 becomes high, and when input lines are 01, 10, 11 then the output pins D1, D2, D3 are high respectively.

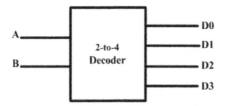

Figure 4.21 Block diagram of 2-to-4 decoder.

Table 4.8 2-to-4 decoder

Inputs		Outputs			
A	B	D0	D1	D2	D3
0	0	1	0	0	0
0	1	0	1	0	0
1	0	0	0	1	0
1	1	0	0	0	1

Based on the observations given in Table 4.8, the Boolean expression of a 2-to-4 decoder is defined as follows:

$$D0 = A'B'$$
$$D1 = A'B$$
$$D2 = AB'$$
$$D3 = AB$$

Circuit diagram

In the implementation of a 2-to-4 decoder, two NOT gates along with four 2-input AND gates are required. Figure 4.22 presents a 2-to-4 line decoder logic circuit.

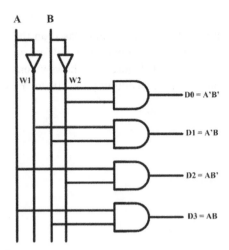

Figure 4.22   Logic circuit of a 2-to-4 decoder.

Verilog Code:
In Verilog code of a 2-to-4 decoder using gate-level modeling, inputs and outputs are defined inside the module Dec_2to4. Inside the module, two NOT gates, as well as four AND gates along with two wires are used.

> **module** Dec_2to4(A,B,D0,D1,D2,D3);
> **input** A,B;
> **output** D0,D1,D2,D3;
> **wire** w1,w2;
> **not** (w1,A);
> **not** (w2,B);
> **and** (D0,w1,w2);
> **and** (D1,w1,B);
> **and** (D2,A,w2);
> **and** (D3,A,B);
> **endmodule**

4.3.3.2   Encoder
The encoder circuit is a combinational circuit with a 2^n input gate with n number of output gates. If 4 is the number of inputs, then 2 are the outputs.

4-to-2 Encoder

In a 4-to-2 encoder, there are four inputs and two output lines. Based on the selection of input lines, outputs are displayed at the output terminal. Based on four input lines, possible outputs are 00, 01, 10, 11. Figure 4.23 shows where four input lines W0, W1, W2, and W3 along with two output pins, Y0 and Y1 are present.

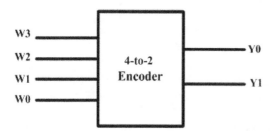

Figure 4.23 Block diagram of 4-to-2 encoder.

Table 4.9 4-to-2 encoder.

Inputs				Outputs	
W3	W2	W1	W0	Y1	Y0
0	0	0	1	0	0
0	0	1	0	0	1
0	1	0	0	1	0
1	0	0	0	1	1

Based on the observations given in Table 4.9, the Boolean expression for the 4-to-2 encoder is defined as follows:

$$Y1 = W2 + W3$$
$$Y0 = W1 + W3$$

Circuit diagram

In the implementation of a 4-to-2 encoder, two OR gates are required as shown in Figure 4.24.

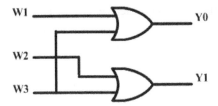

Figure 4.24 Logic circuit of 4-to-2 encoder.

Verilog Code

In Verilog code of a 4-to-2 encoder using gate-level modeling, inputs and outputs are defined inside the module Enc_4to2. Inside the module, two OR gates are used.

```
module Enc_4to2(w1,w2,w3,y0,y1);
    input w1,w2,w3,w4;
    output y0,y1;
    or O_1(y0,w1,w3);
    or O_2(Y1,w2,w3);
endmodule
```

4.3.4 Comparator

The comparator circuit is a combinational circuit which always has three outputs. If two inputs are present in a comparator, the three outputs are $A < B$, $A = B$, and $A > B$. In a 1-bit magnitude comparator, two inputs and three outputs are present.

Table 4.10 1-bit magnitude comparator.

Inputs		Outputs		
A	B	A < B	A = B	A > B
0	0	0	1	0
0	1	1	0	0
1	0	0	0	1
1	1	0	1	0

Based on the observations given in Table 4.10, the Boolean expression for the 1-bit comparator is defined as follows:

A < B = > A'B
A > B = > AB'
A = B = > A'B' + AB

Circuit diagram

In the implementation of a 1-bit comparator circuit, two NOT gates and two AND gates, as well as one XOR gate, are required. This circuit diagram is shown in Figure 4.25.

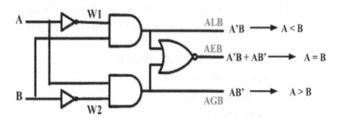

Figure 4.25 Logic circuit of a 1-bit magnitude comparator.

Verilog Code:

In the Verilog code of a 1-bit magnitude comparator circuit using gate-level modeling, inputs and outputs are defined inside the module Com_1. Inside the module, two NOT gates and two AND gates, as well as one XOR gate, along with two wires are used.

```
module Com_1(A, B, ALB, AGB, AEB);
    input A,B;
    output ALB,AGB,AEB;
    wire w1,w2;
    not (w1,A);
    not (w2,B);
    and (ALB,w1,B);
    and (AGB,w2,A);
    nor (AEB,ALB,AGB);
endmodule
```

Review Questions

Q1 Describe the 4-bit adder circuit and write a Verilog code using gate-level modeling.

Q2 Draw the circuit of an 8 × 1 multiplexer using a 2 × 1 multiplexer and write the Verilog code using gate-level modeling.

Q3 Write a Verilog code to construct a 3-to-8 decoder circuit using gate-level modeling.

Q4 Explain the principle of magnitude comparator and write a Verilog code for a 2-bit magnitude comparator using gate-level modeling.

Q5 Write a Verilog code for a 4-bit adder/subtractor circuit using gate-level modeling.

Multiple Choice Questions

Q1 Verilog programming is case sensitive?
  A True
  B False

Q2 RTL in Verilog stands for?
  A Register transfer logic
  B Register transistor logic
  C Register transfer level
  D All

Q3 How many wires are used in a 4 × 1 multiplexer using gate-level modeling?
  A 3
  B 4
  C 5
  D 6

Q4 How many AND gates are called for a 2-to-4 decoder in gate-level modeling?
  A 3
  B 4
  C 6
  D 5

Q5 Which keyword is used for equal operator at the output port of a 1-bit magnitude comparator?

A AND

B OR

C NOR

D NAND

References

[1] Weste, N.H.E, Harris, D., and Banerjee, A. (2011). *CMOS VLSI Design: A Circuits and Systems Perspective*, 3 ed. Pearson Education.

[2] Bhaskar, J. (2015). *A Verilog Primer*. Pearson Education.

[3] Mano, M.M. (1982). *Computer System Architecture*. Prentice Hall.

5

Programming Techniques in Verilog II

5.1 Programming Techniques in Verilog II

Verilog is a Hardware Description language (HDL) used to illustrate a digital system such as a microprocessor, flip-flops (F/Fs), network switch, memory, etc. The Verilog language can be used to describe any digital hardware at any level. The circuit designs developed using HDL are not dependent on technology, are more helpful than schematics and are very simple for debugging and designing, especially for huge circuits. In this chapter, the dataflow model of the circuit is described with the help of different types of circuits.

5.2 Dataflow Model of Circuits

Dataflow modeling does not explain the combinational circuits logic gate. It describes the Boolean funtion of output variable in terms of input variables using operators available in the Verilog library. Here, the data flows from register to register. It requires lesser design steps in comparison to the gate-level model. A number of operators are used by dataflow modeling for the operator to produce the desired results. The model makes use of continuous assignments with the keyword "**assign**."

HDL Operators for the dataflow model include:

- + Binary addition
- − Binary subtraction
- & Bit – wise AND
- | Bit – wise OR

Digital VLSI Design and Simulation with Verilog, First Edition. Suman Lata Tripathi, Sobhit Saxena, Sanjeet Kumar Sinha, and Govind Singh Patel.
© 2022 John Wiley & Sons Ltd. Published 2022 by John Wiley & Sons Ltd.

- ^ Bit – wise XOR
- ~ Bit – wise NOT
- ?: Conditional

5.3 Dataflow Model of Combinational Circuits

Adder and Subtractor
Multiplexer
Decoder
Comparator

5.3.1 Adder and Subtractor

5.3.1.1 Half Adder

In a half-adder circuit, X and Y are the inputs, however S and C are the outputs. For two inputs, there are four possibilities: 00, 01, 10, 11. All four possible combinations, as well as corresponding outputs, are described in Table 5.1 [1].

Table 5.1 Half adder.

Inputs		Outputs	
X	Y	S	C
0	0	0	0
0	1	1	0
1	0	1	0
1	1	0	1

Dataflow description of the half adder
Based on the observations given in Table 5.1, the Boolean expression for a half adder is defined as follows:

$$S = X'Y + XY' = X \oplus Y \text{ and } C = XY$$

Verilog Code
In Verilog code of half adder using dataflow modeling, inputs and outputs are defined inside the module Half_Add. Inside the module, two logic continuous assignments, keyword **assign,** are called (2).

```
moduleHalf_Add(X,Y,S,C);
input X,Y;
output S,C;
assign S = X^Y;
assign C =X & Y;
endmodule
```

5.3.1.2 Half Subtractor

In a half-adder circuit, X and Y are the inputs, whereas D and B are considered outputs. Since there are only two inputs, the input combinations are four: 00, 01, 10, 11. All four possible combinations, as well as corresponding outputs, are described in Table 5.2 (3).

Table 5.2   Half subtractor.

Inputs		Outputs	
X	Y	D	B
0	0	0	0
0	1	1	1
1	0	1	0
1	1	0	0

Based on the observations given in Table 5.2, the Boolean expression for a half subtractor is defined as follows:

The difference, $D = XY' + X'Y$ and borrow, $B = X'Y$

Verilog Code

In Verilog code of the half subtractor using dataflow modeling, inputs and outputs are defined inside the module Half_Subtractor. Inside the module, continuous assignment keyword **assign** is called two times.

```
moduleHalf_Subtractor(X,Y, D,B);
input X,Y;
output D,B;
assign D = X^Y;
assign B = ~X & Y;
endmodule
```

5.3.2 Multiplexer

5.3.2.1 2 × 1 Multiplexer

In the 2 × 1 multiplexer (Figure 5.1), there are two inputs, one select line and one output line. Based on the select line, inputs are reflected in terms of output at the output terminal. As only one select line is present in a 2 × 1 multiplexer, there are two options; 0 and 1.

As shown in Figure 5.1, when the select line becomes 0, the I0 is reflected at output port Y, and when the select line becomes 1, then I1 is considered as output as shown in Table 5.4.

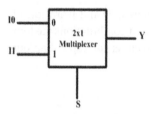

Figure 5.1 Block diagram of 2 × 1 multiplexer.

Table 5.3 2 × 1 multiplexer.

Select	Output
S	Y
0	10
1	11

Based on the observations given in Table 5.3, the Boolean expression for a 2 × 1 multiplexer is defined as follows:

$$Y = S'I0 + SI1$$

Verilog Code

In Verilog code of the 2 × 1 multiplexer using dataflow modeling, inputs and outputs are defined inside the module mux_2x1_df. Inside the module, one continuous assignment keyword **assign** is used.

```
module mux_2x1_df (I0, I1, S, Y);
inputI0, I1, S;
output Y;
assign Y = (~S & I0) 1 (S & I1);
endmodule
```

5.3.2.1 4 × 1 Multiplexer

In the 4 × 1 multiplexer, there are four inputs, two select lines and one output line. Based on the select lines, the inputs are reflected in terms of output at the

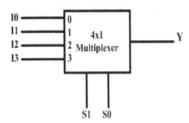

Figure 5.2 Block diagram of 4 × 1 multiplexer.

Table 5.4 4 × 1 multiplexer.

Select lines		Output
S1	S0	Y
0	0	I0
0	1	I1
1	0	I2
1	1	I3

output terminal. As only two select lines are present in a 4 × 1 multiplexer (Figure 5.2), there are four options: 00, 01, 10, 11. As shown in Figure 5.2, when the select line becomes 00, the I0 is reflected at the output port Y, and when the select line becomes 11, then I3 is considered as an output [3].

Based on the observations given in Table 5.4, the Boolean expression for a 4 × 1 multiplexer is defined as follows:

$$Y = S1'S0'I0 + S1'S0I1 + S1S0'I2 + S1S0I3$$

Verilog Code

In Verilog code of a 4 × 1 multiplexer using dataflow modeling, inputs and outputs are defined inside the module mux_4x1. Inside the module, one continuous assignment keyword **assign** is used according to the Boolean expression (2).

```
module mux_4x1(y, i0,i1,i2,i3, s0, s1);
input i0, i1, i2, i3, s0, s1;
output y;
assign y = (~s0 & ~s1 & i0)
           (s0 & ~s1 & i1) |
           (~s0 & s1 & i2) |
           (s0 & s1 & i3);
endmodule
```

Conditional Operator

A ternary operator is another name for a conditional operator (?:), it uses three operands at a time. When the first is true, the operand at the second is evaluated and when the first is false, the operand at the third is evaluated.

Example of a conditional operator.

assign OUT = select ? A: B;

Table 5.5 2×1 multiplexer.

Select line	Output
S	Y
0	I0
1	I1

2×1 multiplexer using a Conditional Operator

Based on the observations given in Table 5.5, the Boolean expression for a 2×1 multiplexer is defined as follows:

Y = S'I0 + SI1.

The conditional operator is used for Y output. The conditional operator (**?**) suggests output Y is equal to data I1, if the select line S is true otherwise, it is I0 [1].

Verilog Code

In Verilog code of a 2×1 multiplexer using a conditional operator, inputs and outputs are defined inside the module mux_2x1. Inside the module, one continuous assignment keyword **assign** is used. Under the assign statement, a conditional operator is used.

```
module mux_2x1(I0, I1, S, Y);
inputI0, I1, S;
output Y;
assign Y = S?I1 : I0;
endmodule
```

4 × 1 multiplexer using a Conditional Operator:

Table 5.6 4 × 1 multiplexer.

Select lines		Output
S1	S0	Y
0	0	I0
0	1	I1
1	0	I2
1	1	I3

Based on the observations given in Table 5.6, the Boolean expression for a 4 × 1 multiplexer is defined as follows:

$$Y = S1'S0'I0 + S1'S0I1 + S1S0'I2 + S1S0I3$$

Verilog Code
In Verilog code of a 4 × 1 multiplexer using the conditional operator, inputs and outputs are defined inside the module mux_4x1. Inside the module, one continuous assignment keyword **assign** is used. Under the assign statement, three conditional operators are used according to the Boolean expression.

```
module mux_4x1(y, i0,i1,i2,i3, s0, s1);
input i0, i1, i2, i3, s0, s1;
output y;
assign y = s1?(s0 ?I3 : i2):(s0?I1 : i0);
endmodule
```

5.3.3   Decoder

2-to-4 Line Decoder

Table 5.7   2-to-4 decoder.

Inputs		Outputs			
A	B	D0	D1	D2	D3
0	0	1	0	0	0
0	1	0	1	0	0
1	0	0	0	1	0
1	1	0	0	0	1

Based on the observations given in Table 5.7, the Boolean expression for a 2-to-4 decoder is defined as follows:

D0 = A'B'
D1 = A'B
D2 = AB'
D3 = AB

Verilog Code
In Verilog code of a 2-to-4 decoder using dataflow modeling, inputs and outputs are defined inside the module Dec_2to4. Inside the module, four continuous assignment keyword assign are used according to the Boolean expression (2).

```
module Dec_2to4(A, B, D0, D1, D2, D3)
    input A, B;
    output D0, D1, D2, D3;
    assign D0 =~ A & ~ B;
    assign D1 =~ A & B;
    assign D2 = A & ~ B;
    assign D3 = A & B;
endmodule
```

5.3.4 Comparator

1-bit Magnitude Comparator
A comparator circuit is a combinational circuit that always has three outputs. If two inputs are present in a comparator, three outputs are $A < B$, $A = B$, and $A > B$. In a 1-bit magnitude comparator, two inputs and three outputs are present.

Table 5.8   1-bit magnitude comparator.

Inputs		Outputs		
A	B	A < B	A = B	A > B
0	0	0	1	0
0	1	1	0	0
1	0	0	0	1
1	1	0	1	0

Based on the observations given in Table 5.8, the Boolean expression for a 1-bit comparator is defined as follows:

$$A\langle B=\rangle A'B$$
$$A > B => AB'$$
$$A = B => A'B' + AI$$

Verilog Code:
In Verilog code of the 1-bit magnitude comparator using dataflow modeling, inputs and outputs are defined inside the module Com_1. Inside the module, three continuous assignments keyword **assign** are used according to the Boolean expressions.

```
module Com_1(A, B, ALB, AGB, AEB);
input A,B;
output ALB, AGB, AEB;
assign ALB =~ A & B;
assignAEB =~ (A ^ B);
assign AGB = A & ~ B;
endmodule
```

5.4 Testbench

Testbenches are used for simulation purposes. There is no need for physical hardware to design and simulate the testbench. To a Verilog-based design, the testbench defines a sequence of inputs to be applied by the simulator.

The testbench is incorporated into the module to be tested. There is no need for input as well as output nodes for the testbench. The reg and outputs with wire are required to define inputs. The keyword **initial** is used to define the inputs. All the input values are designated between begin and end [2].

The end of the simulation is determined with $finish or $stop.

    #100$stop;
//the simulation will be suspended at time = 100
    #900$finish;
//the simulation will be terminated at time = 1000

Testbench module for a simple circuit

```
moduleckt_tb();
regA, B, C;
wirex, y;
Module_nameCKT(A, B,C, x, y);
initial
begin
A = 1'b0; B = 1'b0; C = 1'b0;
#100
A = 1'b1; B = 1'b1; C = 1'b1;
#100$stop;
end
endmodule
```

5.4.1 Dataflow Model of the Half Adder and Testbench

Box 5.1
//Testbench of Half adder

module HA_tb();
 reg A,B;
 wire S,C;
 Half _Add Half (A, B, S, C);
 initial
 begin
 A =0; B =0;
     # 5A = 0; B =1;
     # 5A = 1; B =0;
     # 5A = 1; B =1;
 #5 $stop;
 end
 endmodule

In a half-adder circuit, an input port has two inputs A and B whereas the output terminal has two outputs S and C. The main module name of the half adder using dataflow modeling is Half-Add [2]. This main module is called inside the testbench module of half adder HA_tb. Since A and B are defined as input ports in the main module, these two variables are defined as **reg** in the testbench module. The out port defined in the main module S and C are declared by wire in the testbench module.

//Half adder module

```
module Half_Add(A, B, S, C)
input A,B;
output S,C;
assign S =A^B;
assign C =A&B;
endmodule
```

With two input variables four possibilities arise, in testbench all possibilities are given a certain time delay while calling. In the testbench of the half adder illustrated in Box 5.1, 5 ns is used for all four possible states.

5.4.2 Dataflow Model of the Half Subtractor and Testbench

In a half-subtractor circuit, the input port has two inputs X and Y, whereas the output terminal has two outputs, D and B. The main module name of a half subtractor using dataflow modeling is Half_S. This main module is called inside the testbench module of the half subtractor HS_tb. Since X and Y are defined as input ports in the main module, these two variables are defined as **reg** in the testbench module. The out port defined in the main module as D and B are declared by wire in the testbench module.

Box 5.2

//Testbench of Half Subtractor

```
module HS_tb();
reg X,Y;
wire D,B;
Half_SHalf(X, Y, D, B);
initial
begin
        X=0; Y =0;
     # 5X =0; Y =1;
     # 5X =1; Y =0;
     # 5X =1; Y =1;
#5 $stop;
end
endmodule
```

//Half Subtractor module

```
module Half_ S(X, Y, D, B);
input X,Y;
output D,B;
assign D=X^Y;
assign B=~X&Y;
endmodule
```

With two input variables, four possibilities arise in the testbench, all possibilities are given a certain time delay while calling. In the testbench of the half adder illustrated above in Box 5.2, 5 ns is used for all four possible states.

5.4.3   Dataflow Model of 2 × 1 Mux and Testbench

In the 2×1 multiplexer circuit, the input port has two inputs, I0 and I1, along with one select line S, whereas the output terminal has one output Y. The main module name of a 2×1 multiplexer using dataflow modeling is mux_2x1_df. This main module is called mux_tb inside the testbench module of a 2×1 multiplexer. Since I0, I1, and S are defined as input ports in the main module, these three variables are defined as **reg** in the testbench module. The output defined in main modules Y, are declared by wire in the testbench module.

Box 5.3

// Testbench of 2 × 1 mux

```
modulemux_tb();
regI0,I1,S;
wireY;
mux_2x1_df Mux(Y, I0, I1, S);
initial
begin
    #5 I0 =1'bl; I1 =1'b0;
    #5 S =1'b0;
    #5 S =1;b1;
    #5 $stop;
end
endmodule
```

//2 × 1 mux module

> module mux_2x1_df (I0, I1, S, Y);
> inputI0, I1, S;
> output Y;
> assign Y = (~ S & I0) 1(S & I1);
> endmodule

//2 × 1 mux module with a Conditional operator

> module mux_2x1_df(I0, I1, S, Y);
> inputI0, I1, S;
> output Y;
> assign Y = S?I1 : I0;
> endmodule

5.4.4 Dataflow Model of 4 × 1 Mux and Testbench

In the 4 × 1 multiplexer circuit, the input port has four inputs: I0, I1, I2, and I3, along with two select lines S0 and S1, whereas the output terminal has one output, Y. The main module name of a 4 × 1 multiplexer using dataflow modeling is mux_4x1_df. This main module is called mux_tb inside the testbench module of a 4 × 1 multiplexer. Since I0, I1, I2, I3, and S0, S1 are defined as input ports in the main module, these six variables are defined as **reg** in the testbench module. The output defined in the main module is Y, which is declared by wire in the testbench module.

//4 × 1 mux module

> module mux_4x1(y, I0, I1,I2, I3, s0, s1);
> input I0, I1, I2, I3, s0, s1;
> output y;
> assign Y = (~ s0 & ~ s1 & I0)|
>   (s0 & ~ s1 & I1)|
>   (~ s0 & s1 & I2)|
>   (s0 & s1 & I3);
> endmodule

Box 5.4

// Testbench of 4 × 1 mux

```
modulemux_tb();
regI0,I1,I2,I3,S1,S0;
wireY;
mux_4xl Mux(Y,I0,I1,I2,I3,S1,S0);
initial
begin
   I0=1; I1=0; I2=1; I3=0;
     # 5 s1 =0; s0=0;
     # 5 S1 =0; s0=1;
     # 5 S1 =1; s0=0;
     # 5 S1 =1; s0=1;
   # 5 $finish;
end
endmodule
```

//4 × 1 mux module with a Conditional operator

```
module mux _ 4x1(y, i0,i1,i2,i3, s0, s1);
input i0, i1, i2, i3, s0, s1;
output y;
assign y = s1?(s0 ?I3 : i2):(s0?I1 : i0);
endmodule
```

5.4.5 Dataflow Model of 2-to-4 Decoder and Testbench

In the 2-to-4 decoder circuit, input port has two inputs A and B whereas the outport has four ports: D0, D1, D2, and D3. The main module name of the 2-to-4 decoder using dataflow modeling is Dec_2to4. This main module is called Dec_tb inside the testbench module of the 2-to-4 decoder. As A and B are defined as input ports in the main module, these two variables are defined as **reg** in the testbench module. The outputs defined in the main module are D0, D1, D2, and D3; those are declared by **wire** in the testbench module.

Box 5.5

// Testbench of 2-to-4 Decoder

```
moduleDec_tb();
reg A,B;
wire D0,D1,D2,D3;
Dec_2to4 Dec(A,B,D0,D1,D2,D3);
initial
begin
    A=0; B=0;
    #5A=0; B=1;
    #5A=1; B=0;
    #5A=1; B=1;
    #5 $finish;
end
endmodule
```

//**2-to-4 Decoder module**

```
module Dec_2to4(A,B,D0,D1,D2,D3);
input A,B;
output D0,D1,D2,D3;
assign D0=~A&~B;
assign D1=~A&B;
assign D2=A&~B;
assign D3=A&B;
endmodule
```

Review Questions

Q1 Write a Verilog code of a 4-bit full adder using dataflow modeling with testbench.

Q2 Implement a 4-bit adder/subtractor circuit using dataflow modeling of Verilog.

Q3 Design a 8×1 multiplexer using the 2×1 multiplexer using the dataflow model.

Q4 Write a Verilog code of a 3-to-8 decoder using the dataflow model and illustrate their testbench.

Q5 Write a Verilog code of 2-bit magnitude comparator using the dataflow model.

Multiple Choice Questions

Q1 Which operator is not used as dataflow modeling?
A +
B –
C @
D &

Q2 In the testbench module, the output ports are declared as?
A reg
B wire
C o/p
D i/p

Q3 For the conditional operator, which keyword is used?
A >
B <
C ?
D &

Q4 Which of the following is the continuous assignment keyword?
A module
B assign
C initial
D finish

Q5 Which keyword is used for negation?
A *
B $
C ~

References

[1] Weste, N.H.E., Harris, D.,, and Banerjee, A. (2004). *CMOS VLSI Design: A Circuits and Systems Perspective*, 3e. Upper Saddle River, NJ: ed. Pearson.

[2] Bhaskar, J. (1999). *A Verilog HDL Primer*. Upper Saddle River, NJ: Pearson.

[3] Mano, M.M. (1992). *Computer System Architecture*. Upper Saddle River, NJ: Pearson.

6

Programming Techniques in Verilog II

6.1 Programming Techniques in Verilog II

The behavioral model of a circuit is a representation of all blocks in the algorithmic level. In the behavioral model, the keyword is always used with different procedural assignments. In this model, the output of the circuit is defined as **reg**. Verilog HDL mainly has four design levels with the behavioral model as the highest level of abstraction. The behavioral model uses algorithm or truth tables for the implementation of digital circuit design. It does not need information about logic block representation [1]. Both combination and sequential circuit designs can be done at the behavioral level. The programming concept in the behavioral model is almost similar to C language.

In the highest level of the behavioral model, each statement execution is performed through a triggering signal. The statement execution is initiated through the keywords *always* or *initial*, called procedural statements. These procedural statements are used to define the activity flow of the particular logic block. The nesting of *always* and *initial* blocks are not allowed in Verilog HDL. The *initial* blocks are mainly used to assign values to input and output variables.

Properties of behavioral statements are:

- Each *initial* block executes only once during the simulation and statements start executions at a zero instant of time.
- For multiple *initial* blocks, the execution starts at zero instant.
- The *always* block also starts at execution at the zero instant but the flow of activities goes in a looping manner.

Digital VLSI Design and Simulation with Verilog, First Edition. Suman Lata Tripathi, Sobhit Saxena, Sanjeet Kumar Sinha, and Govind Singh Patel.
© 2022 John Wiley & Sons Ltd. Published 2022 by John Wiley & Sons Ltd.

The multiple statements inside *always* or *initial* blocks are grouped within the keywords *begin* and *end* [2].

always [timing_control] procedural_statement

Example:

example of a clock clk signal in Verilog

always @posedgeclk

#5 clk v= ~ clk;

A procedural continuous assignment is a procedural statement, that can appear inside an **always statement block** or an **initial statement block**. This is different from a continuous assignment; a continuous assignment occurs **outside** the *initial* or *always* block.

6.2 Behavioral Model of Combinational Circuits

6.2.1 Behavioral Code of a Half Adder Using If-else

A half-adder circuit has two inputs A as well as B whereas the outputs are Sum (S) and Carry (C). For two inputs, there are four possibilities: 00, 01, 10, 11. All possible four combinations, as well as corresponding outputs, are described in Table 6.1 [1].

Table 6.1 Half adder.

Inputs		Outputs	
A	B	S	C
0	0	0	0
0	1	1	0
1	0	1	0
1	1	0	1

Based on the observations of Table 6.1, the Boolean expression for the half adder is defined as follows:

$$S = A'B + AB' = A \oplus B \text{ and } C = AB$$

Verilog Code

In Verilog code of the half adder using behavioral modeling, inputs and outputs are defined inside the module half adder. Inside the module, the if-else statement is called according to the expression of half-adder outputs [2].

```
modulehalfadder(S,C,A,B);
input A, B;
output S, C;
reg S,C;
always@(A or B)
begin
if (A = = 0&& B = = 0)
begin S = 0; C = 0;
end
elseif (A = = 1&& B = = 1)
begin S = 0; C = 1;
end
else
begin S = 1; C = 0;
end
end
endmodule
```

6.2.2 Behavioral Code of a Full Adder Using Half Adders

In the full-adder circuit A, B, and C are considered as three inputs whereas S and C are considered as output. Since there are only three inputs, the input combinations are eight: 000, 001, 010, 011, 100, 101, 110, 111. All possible four combinations, as well as corresponding outputs, are described in Table 6.2.

Table 6.2 Full adder.

Inputs			Outputs	
A	B	Cin	S	Co
0	0	0	0	0
0	0	1	1	0
0	1	0	1	0
0	1	1	0	1
1	0	0	1	0
1	0	1	0	1
1	1	0	0	1
1	1	1	1	1

Based on the observations of Table 6.2, one 3-input XOR gate, along with three 2-input AND gates, as well as one 3-input OR gate are required for the circuit design of the full adder. The expression in terms of three input variables of a full adder (Figure 6.1) is defined as follows:

$$S = A'B'Cin + A'BCin' + AB'Cin' + ABCin = A \oplus B \oplus Cin$$
$$\text{and } Co = A'BCin + AB'Cin + ABCin' + ABCin = AB + BCin + ACin$$

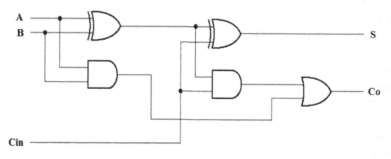

Figure 6.1 Logic circuit of a full adder [1].

Verilog Code

Implementation of the full adder with half adders requires two half adders as well as one OR gate. In Verilog code of the full adder with half adders, a two half-adder module is called inside the full adder main module along with one OR gate [2].

```
module fulladder (S,C,x,y,z);
input x,y,z;
output S,C;
reg S,C;
wire S1,D1,D2;
//Instantiate the half adders
halfadder HA1(S1,D1,x,y),
HA2(S,D2,S1,z);
or g1(C,D2,D1);
endmodule
```

6.2.3 Behavioral Code of a 4-bit Full Adder (FA)

Figure 6.2 presents a logic block of the 4-bit full adder.

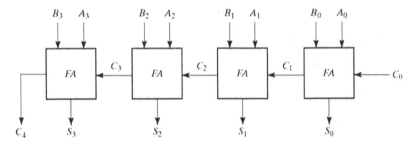

Figure 6.2 Block diagram of a 4-bit full adder [1].

Verilog Code

module 4bit_adder (S,C4,A,B,C0);
input [3:0] A,B;
input C0;
output [3:0] S;
output C4;
reg S,C4;
wire C1,C2,C3;
//Instantiate the full adder
fulladder FA0 (S[0],C1,A[0],B[0],C0),
FA1 (S[1],C2,A[1],B[1],C1),
FA2 (S[2],C3,A[2],B[2],C2),
FA3 (S[3],C4,A[3],B[3],C3);
endmodule

6.2.4 Behavioral Model of Multiplexer Circuits

A multiplexer is another combinational circuit which is an important part of digital electronics. Multiplexers are known as data-selector devices, which means they select only one output at a time out of the many available inputs, based upon the select lines [1].

6.2.4.1 Behavioral Code of a 2 × 1 Multiplexer

In a 2 × 1 multiplexer, there are two inputs, one select line and one output line. Based on the select line, the inputs are reflected in terms of output at output terminal. As only one select line is present in a 2 × 1 multiplexer, there are two options 0 and 1. As shown in the block diagram (Table 6.3), when 0 is selected at the select line, the A is reflected at the output port out, and when the select line becomes 1, then B is considered as output [3].

Table 6.3 2 × 1 multiplexer.

Select line	Output
select	OUT
0	A
1	B

***Verilog Code*: 2x1 multiplexer using if else statement**

```
module mux2x1_bh(A,B,select,OUT);
input A, B, select;
output OUT;
reg OUT;
always@(select or A or B)
if (select = = 1) OUT = B;
else OUT = A;
endmodule
```

***Verilog Code*: 2x1 multiplexer using case statement**

```
module mux2x1(A,B,select,out);
input A,B;
input select;
output out;
reg out;
always @(A or B or select)
case (select)
1'b0: out = A;
1'b1: out = B;
endcase
endmodule
```

6.2.4.2 Behavioral Code of a 4 × 1 Multiplexer

In the 4 × 1 multiplexer, there are four inputs, two select lines and one output line. Based on the select lines, the inputs are reflected in terms of output at an output terminal. Since only two select lines are present in the 4 × 1 multiplexer, there are four options: 00, 01, 10, 11. As shown in Figure 6.3 representing a block diagram of the 4 × 1 multiplexer (Table 6.4), when the select line becomes 00 the I0 is reflected at output port Y, and when the select line becomes 11, then I3 is considered as output [3].

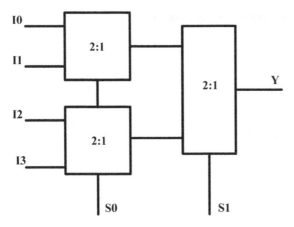

Figure 6.3   Logic circuit of 4 × 1 multiplexer.

Table 6.4   4 × 1 multiplexer.

S1	S0	Y
0	0	I0
0	1	I1
1	0	I2
1	1	I3

Verilog Code: **4x1 multiplexer using 2 × 1 multiplexer**

```
//4X1 multiplexer//
module Mux_4X1(S,I,Y);
input [3:0]I;
input [1:0] S;
output Y;
reg Y;
wire C1,C2;
//Instantiate 2 × 1 multiplexer
mux2x1_bh M0 (I[0],I[1], S[0],C1),
M1 (I[2],I[3], S[0],C2),
M2 (C1,C2,S[1], Y);
endmodule
```

Verilog Code: **4x1 multiplexer using case statement**

```
module mux4x1(i0,i1,i2,i3,S,y);
input i0,i1,i2,i3;
input [1:0] S;
output y;
reg y;
always @(i0 or i1 or i2 or i3 or S)
case (S)
2'b00: y = i0;
2'b01: y = i1;
2'b10: y = i2;
2'b11: y = i3;
endcase
endmodule
```

6.2.5 Behavioral Model of a 2-to-4 Decoder

In a 2-to-4 decoder, there are two input lines as well as four output lines. Based on the selection of input lines, outputs are displayed at the output terminal. Since two input lines are present, so there are a total of four possibilities: 00, 01, 10, 11. A block diagram (Figure 6.4) is shown where, when the input lines become 00, the output pin D0 becomes high, and when input lines are 01, 10, 11 then output pins D1, D2, D3 are high respectively (Table 6.5).

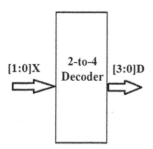

Figure 6.4 Block diagram of a 2-to-4 decoder.

Table 6.5 2-to-4 decoder.

Inputs		Outputs			
X1	X0	D0	D1	D2	D3
0	0	1	0	0	0
0	1	0	1	0	0
1	0	0	0	1	0
1	1	0	0	0	1

Verilog Code: Behavioral code of a 2-to-4 decoder using if else statement

```
module Dec_2to4(X, D);
input [1:0]X;;
output [3:0]D;
reg D;
always @(X)
begin
if(X = = 2'b00) D = 4'b0001;
else if (X = = 2'b01) D = 4'b0010;
else if (X = = 2'b10) D = 4'b0100;
else (X = = 2'b11) D = 4'b1000;
end
endmodule
```

Verilog Code: Behavioral code of a 2-to-4 decoder with case statement

In the implementation of a 2-to-4 decoder with case statement, a case X is required. Here, X is the input variable. As four posible input combinations are present, when 00 is selected at the input side, D0 is reflected at output. In same way, when 01, 10, 11 are selected at input ports, then D1, D2, D3 are considered at the output port respectively.

```
module Dec_2to4(X, D);
input [1:0]X;;
output [3:0]D;
reg D;
always @(X)
begin
```

```
case (X)
2'b00: D = 4'b0001;
2'b01: D = 4'b0010;
2'b10: D = 4'b0100;
2'b11: D = 4'b1000;
endcase
end
endmodule
```

6.2.6 Behavioral Model of a 4-to-2 Encoder

In a 4-to-2 encoder, there are four input lines, as well as two output lines. Based on the selection of input lines, outputs are displayed at the output terminal. Four input lines are 0001, 0010, 0100, 1000 which are present at the input side. A block diagram is shown in Figure 6.5 where, when the input line becomes 0001, the output pin becomes 00, and when input lines are 0010, 0100, 1000 then outputs are 01, 10, and 11 respectively (Table 6.6).

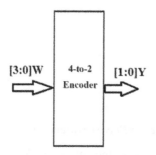

Figure 6.5   Block diagram of a 4-to-2 decoder.

Table 6.6   Decoder truth table.

Inputs				Outputs	
W3	**W2**	**W1**	**W0**	**Y0**	**Y1**
0	0	0	1	0	0
0	0	1	0	0	1
0	1	0	0	1	0
1	0	0	0	1	1

Verilog Code: Behavioral code of a 4-to-2 encoder using if else statement

```
module Enc_4to2(W, Y);
input [3:0]W;;
output [1:0]Y;
reg Y;
always @(W)
begin
if(W = = 4'b0001) Y = 2b'00;
else if (W = = 4'b0010) Y = 2b'01;
else if (W = = 4'b0100) Y = 2b'10;
else (W = = 4'b1000) Y = 2b'11;
end
endmodule
```

Verilog Code: Behavioral code of a 4-to-2 encoder with case statement

In the implementation of a 4-to-2 encoder with case statement, a case W is required. Here, W is an input variable with a size of 4 bits. As four inputs are present, when 0001 is selected at input, side 00 is reflected at output. In the same way, when 0010, 0100, and 1000 are selected at input ports then 01, 10, 11 are considered at output ports respectively.

```
module Enc_4to2(W, Y);
input [3:0]W;;
output [1:0]Y;
reg Y;
always @(W)
begin

case (W)

4'b0001: Y = 2'b00;
4'b0010: Y = 2'b01;
4'b0100: Y = 2'b10;
4'b1000: Y = 2'b11;
endcase
end
endmodule
```

6.3 Behavioral Model of Sequential Circuits

In the sequential type of digital logic circuit, the output depends on both present state inputs and past outputs of the circuit. A sequential circuit has memory elements which store the outputs of the circuit.

6.3.1 Behavioral Modeling of the D-Latch

D-Latch (Figure 6.6) is a basic element of a sequential circuit which has a single input D along with the enable signal. According to the enable signal, the input is reflecting at output port Q.

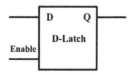

Figure 6.6 Block diagram of the D-Latch.

Verilog code of the D-Latch:

In Verilog code of the D-Latch, *always* is used for the enable signal as well as the input signal d. If the enable signal is there, the output q becomes the same as input d.

```
moduled_latch(d,en,q);
input d, en;
output q;
reg q;
always@(en,d)
begin
if (en)
q ≤ d;
end
endmodule
```

D-LatchTestbench:

```
module d-tb();
regd,en;
wire q;
```

d_latchD1(d,en,q);

Initial

begin

en = 1'b0;

d = 1'b1;

#40 $finish;

end

always #5 en = ~ en;

always #7 d = ~ d;

endmodule

6.3.2   Behavioral Modeling of the D-F/F

D-F/F (Figure 6.7) is a type of flip-flop (F/F) which can store one bit of data that has a single input D as well as a single clock signal. According to the clock signal, the input is reflecting at output port Q. The clock signal may be either positive or negative.

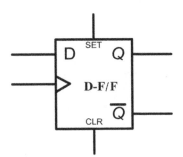

Figure 6.7   Block diagram of a D-F/F.

Table 6.7   D-F/F truth table.

D	Q(next)
0	0
1	1

Verilog code of D-F/F:

In Verilog code of a D-F/F, *always* is used for the clock signal, reset and as input signal d. If the clock signal is there, the output q becomes the same as input d (Table 6.7).

```
moduled_ff(clk,d,rst,q);
inputclk,d,rst;
output q;
reg q;
always@(posedgeclk,negedgerst)
begin
if (rst)
q ≤ 1'b0;
else
q ≤ d;
end
endmodule
```

D-FF Testbench:

```
module d-tb();
regclk,d,rst;
wire q;
d_ffD1(clk,d,rst,q);
Initial
begin
clk = 1'b0; rst = 1'b0; d = 1'b0;
#5 rst = 1'b1;
#7 rst = 1'b0;
#13 rst = 1'b1;
#40 $finish;
end
always #3 clk = ~ clk;
always #5 d = ~ d;
endmodule
```

6.3.3 Behavioral Modeling of the J-K F/F

J-K F/F (Figure 6.8) is another type of F/F which stores a single bit of data that has two inputs, J and K along with the clock signal. According to the clock pulse, the input reflects at output port Q. The clock signal may be either positive or negative.

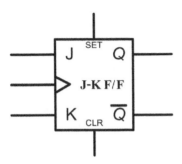

Figure 6.8 Block diagram of the J-K F/F.

Table 6.8 J-K F/F.

J	K	Q (next)
0	0	Q
0	1	0
1	0	1
1	1	Q'

Verilog code of J-K F/F:

In Verilog code of a J-K F/F, an *always* statement is used for the clock signal, as well as the input signals J and K. If the clock signal is there, the output q depends on the input combination of J and K (Table 6.8).

```
module JK (q, q1, j, k, clk);
input j, k, clk;
output q, q1;
reg q, q1;
Initial
begin
q = 1'b0; q1 = 1'b1;
end
always @ (posedgeclk)
begin
```

```
case ({j,k})
{1'b0,1'b0}: begin q = q; q1 = q1; end
{1'b0,1'b1}: begin q = 1'b0; q1 = 1'b1; end
{1'b1,1'b0}: begin q = 1'b1; q1 = 1'b0; end
{1'b1,1'b1}: begin q = ~ q; q1 = ~ q1; end
endcase
end
endmodule
```

J-K F/F Testbench:

```
moduleJK_tb();
regclk;
reg j;
reg k;
wire q, q1;
JK J1(q,q1,j,k,clk);
initial
begin
Clk = 1'b0; j = 1'b1; k = 1'b1;
#25 $finish;
end
always #2 clk = ~ clk;
endmodule
```

6.3.4   Behavioral Modeling of the D-F/F Using J-K F/F

In the implementation of a D-F/F by using J-K F/F as shown in Figure 6.9, one NOT gate is used between the inputs J and K. According to the clock signal, the input reflects at output port Q. The clock signal may be either positive or negative.

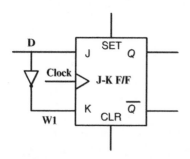

Figure 6.9   Block diagram of the D-F/F using J-K F/F.

Verilog code:

In Verilog code of a D-F/F using J-K F/F, the J-K F/F module is called along with the NOT gate. J-K F/F uses one *always* statement for the clock signal, as well as an input signal J and K. If the clock signal is there, the output q depends on the input combination of J and K [2].

```
moduled_ff(clk, d, q, q1);
inputclk,d;
output q, q1;
reg q, q1;
wire w1;
not N_1 (w1,d);
JK JK_d(q, q1, d, w1, clk);
endmodule
```

6.3.5 Behavioral Modeling of the T-F/F Using J-K F/F

In the implementation of a T-F/F using J-K F/F as shown in Figure 6.10, both inputs J and K are shorted to each other. According to the clock signal, the input reflects at output port Q. The clock signal may be either positive or negative [2].

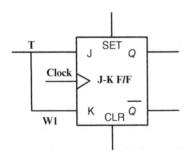

Figure 6.10 Block diagram of the J-K F/F using T-F/F.

Verilog code:

In Verilog code of a T-F/F using J-K F/F, the J-K F/F module is called as both inputs are same as T. J-K F/F uses one *always* statement for the clock signal, as well as the input signals J and K. If the clock signal is there, the output q depends on the input combination of J and K.

```
module T_ff(clk, d, q, q1);
input clk,T;
output q, q1;
reg q, q1;
JK JK_d(q, q1, T, T, clk);
endmodule
```

6.3.6 Behavior Modeling of an S-R F/F Using J-K F/F

In the implementation of the S-R FF using J-K FF as shown in Figure 6.11, both inputs J and K are considered as S and R. According to the clock signal, the input reflects at output port Q. The clock signal may be either positive or negative [3].

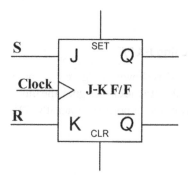

Figure 6.11 Block diagram of an S-R F/F using J-K F/F.

Verilog code:

In Verilog code of an S-R F/F using J-K F/F, the J-K F/F module is called as both inputs J and K are the same as S and R. J-K F/F uses one *always* statement for the clock signal, as well as the input signal J and K. If the clock signal is there, the output q depends on the input combination of J and K [2].

```
module SR_ff(clk, S, R, q, q1);
input S, R;
input clk;
output q, q1;
reg q, q1;
JK JK_SR(q, q1, S, R, clk);
endmodule
```

Review Questions

Q1 Illustrate a clock generator with always statement in Verilog.

Q2 Write a behavioral code of a blocking statement.

Q3 Write a behavioral code for an 8 × 1 multiplexer in Verilog.

Q4 Write a Verilog code for a 4 × 1 multiplexer with case statement.

Q5 Write a Verilog program of a 4-bit counter by using behavioral modeling.

Multiple Choice Questions

Q1 The output of J-K flip-flop when J = 1, K = 0 is
  A 1
  B 0
  C 11
  D 10

Q2 For J-K flip-flop with J = 1, K = 0, the output after clock pulse will be
  A 0
  B 1
  C High impedance
  D No change

Q3 The truth table for an S-R flip-flop has how many VALID entries?
  A 2
  B 3
  C 4
  D 1

Q4 A NOT gate placed between the S and R input of S-R flip-flop results in
  A S-R F/F
  B J-K F/F
  C D-F/F
  D T-F/F

Q5 The logic circuits whose outputs at any instant of time depend not only on the present input but also on the past outputs are called
  A Combinational circuits
  B Sequential circuits
  C Arithmetic circuits
  D None

References

[1] Weste, N.H.E., Harris, D., and Banerjee, A. (2011). *CMOS VLSI Design: A Circuits and Systems Perspective*, 3e. Upper Saddle River, NJ: ed. Pearson.

[2] Bhaskar, J. (2015). *A Verilog HDL Primer*. Upper Saddle River, NJ: Pearson.

[3] Mano, M.M. (1982). *Computer System Architecture*. Upper Saddle River, NJ: Pearson.

7

Digital Design Using Switches

The switch-level model plays an important role in any prefabrication design and analysis in obtaining the knowledge about the actual number of hardware components and interconnections as well as placing the transistor as per the functionality. In comparison to other levels of abstraction in Verilog models, the switch-level model is quite complex and the circuit is designed at the transistor level. For the design at the switch level, the designer must know about design functionality, hardware components and their interconnections. Doing VLSI design at this level also supports prefabrication fault analysis to obtain the additional hardware design for testing. Since this is the lowest level of abstraction among all other Verilog models, the designer needs knowledge of functional logic block and interconnections. In a switch-level model, the design can be done using switches such as the p-channel metal–oxide–semiconductor field-effect transistor MOSFET(PMOS), n-channel MOSFET (NMOS) and transmission gate (TG) or the Complementary MOSFET (CMOS). This chapter includes digital circuit design at the level of switches using Verilog coding known as the switch-level model.

7.1   Switch-Level Model

Verilog HDL supports the hardware design at the level of transistor from the knowledge of design functionality and interconnections. Such transistor level hardware design can be performed with the help of switches like NMOS, PMOS, CMOS, and TG, etc. [1–2]. The requirement of low-power consumption and higher IC packing density leads the designer to explore new MOSFET architectures with low-leakage current and operating voltages. Since most of

Digital VLSI Design and Simulation with Verilog, First Edition. Suman Lata Tripathi, Sobhit Saxena, Sanjeet Kumar Sinha, and Govind Singh Patel.
© 2022 John Wiley & Sons Ltd. Published 2022 by John Wiley & Sons Ltd.

the IC are made of MOSFETs, the understanding of switches based on MOSFETs becomes essential. The understanding of CMOS technology also plays an important role in any digital circuit design at transistor level.

7.2 Digital Design Using CMOS Technology

Most of the ICs are designed with CMOS technology where both the n-channel MOSFET and p-channel MOSFET are used to implement any functionality. CMOS technology is preferred. MOSFET is a four-terminal device. The four terminals are source, drain, gate, and body. The body is connected to the source terminal thus making it a three-terminal device. MOSFET is used for applications such as switching and amplification of electronic signals. Voltage is applied at the gate terminal which controls the flows of current from source to drain. The source is the terminal where the majority of carriers enter. The drain terminal collects the majority of charge carriers from the source. The gate terminal allows the flow of charge carriers from the source to drain when the voltage is applied to it. It can be categorized as: n-channel MOSFET(NMOS) and p-channel MOSFET(PMOS). The channel conduction is due to electrons and holes in NMOS and PMOS, respectively. The higher mobility of electrons lead to NMOS as a suitable choice in comparison to PMOS which has lower hole mobility. In a CMOS-based digital design, NMOS and PMOS are used parallelly to obtain the logic transitions either from 0-1 or 1-0. Also, the NMOS and PMOS are equal in number for the implementation of any digital design using CMOS technology. The number of transistors depend on the number of the input variable. Total number transistors:

$$N_T \geq 2n$$

Where n is the number of input variables. Figure 7.1 shows CMOS design including the pull-up and pull-down networks.

Pull-Up Network
The pull-up network is responsible for the output node transitions from logic 0 to logic 1. It is made from a few PMOS, depending on functionality.

Pull-Down Network
The pull-down network is responsible for the output node transitions from logic 1 to logic 0. It is made from a few NMOS, depending on the functionality.

For, $F = \overline{A.B.C...}$

NMOS will be series connected in a pull-down network and PMOS will, in parallel, be connected in a pull-up network.

For, $F = \overline{(A + B + C \ldots)}$

NMOS will be connected in parallel in the pull-down network and PMOS will be series connected in a pull-up network.

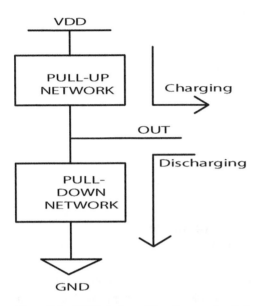

Figure 7.1 CMOS design with pull-up and pull-down network.

7.3 CMOS Inverter

A basic CMOS inverter consists of two transistors including one NMOS and one PMOS. Here, PMOS and NMOS have preferred logic 0-1 and logic 1-0 for perfect logic transitions. For input at logic 0, PMOS transistors will conduct and NMOS remains in an OFF-state. The output node starts charging through the supply (V_{DD}) to maintain logic 1. For input at logic 1, the NMOS transistor will conduct and PMOS goes in to an OFF-state. So, for input at logic 1, the output node starts discharging through ground (GND) to maintain logic 0. This way, an inverter provides an output which is a complement to input function. Here, PMOS and NMOS are working as pull-up and pull-down transistors.

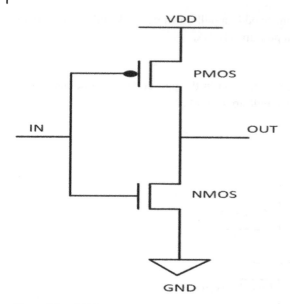

Figure 7.2 CMOS inverter.

7.4 Design and Implementation of the Combinational Circuit Using Switches

Verilog consists of a library switch that can be used for logic implementation at the level of the transistor. This kind of implementation has proved useful as a means to gain knowledge about the internal connections and for placing transistors for a particular logic implementation.

7.4.1 Types of Switches

MOS switches can be categorized as two switches; NMOS and PMOS. The switches NMOS and PMOS can be instantiated using Verilog primitives *nmos* and *pmos* respectively. Figure 7.3 shows the symbols of NMOS and PMOS switches.

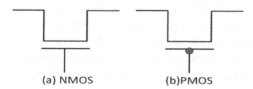

(a) NMOS   (b)PMOS

Figure 7.3 MOS switches (a) NMOS (b) PMOS.

Example:

nmos n1 (drain, source, gate);   *//syntax for NMOS switch instantiation*

pmos p1(drain, source, gate);   *//syntax for PMOS switch instantiation*

Similar to logic gates, switches can also be presented as Verilog primitives. The instantiation of Verilog primitive with an instant name is optional as defined here as n1 and p1.

7.4.2 CMOS Switches

CMOS is another example of switches that are declared with the keyword *cmos* in Verilog HDL. A CMOS device can be designed with an integrated NMOS and PMOS device. Figure 7.4 presents the symbol of a CMOS switch.

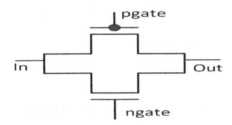

Figure 7.4   Symbol of a CMOS switch.

Example:

cmos c1(Out, In, ngate, pgate);   *//syntax for CMOS switch instantiation*

7.4.3 Resistive Switches

MOS, CMOS, and *bidirectional switches* can also be designed as corresponding *resistive* switches. Resistive switches are known as high-impedance (source-drain) switches while regular switches have been considered as low-impedance (source-drain) switches. These switches are declared and prefixed with 'r' to the corresponding regular switch as a Verilog primitive. Figure 7.5 shows the symbols of resistive switches.

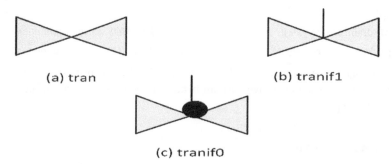

(a) tran   (b) tranif1

(c) tranif0

Figure 7.5   Resistive Switches (a) tran (b) tranif1 (c) tranif0.

Keywords of resistive switches are:
rnmos   *//resistive NMOS switch*
rpmos   *//resistive PMOS switch*
rcmos   *//resistive CMOS switch*
rtran   *// bidirectional resistive switch without control.*
rtranif1   *// bidirectional resistive switch with control*
rtranif0   *// bidirectional resistive switch with control_bar*

7.4.4   Bidirectional Switches

In NMOS, PMOS, and CMOS switches, the current flows from drain to source. Sometimes, design requirements also need devices that can conduct in both directions. In bidirectional switches, the data can be processed on both sides of the device i.e., from drain to source or source to drain. Verilog primitives also support bidirectional switches that are *tran, tranif0,* and *tranif1* without control, with inverted control and control terminals respectively.

Example:
tran x1(port1, port2);   *//instantiation of tran as a switch with name x1*
tranif0 x2(port1, port2, control); *//instantiation of tranif0 as a switch with name x2*
tranif1 x3(port1, port2, control); *//instantiation of tranif0 as a switch with name x2*

7.4.5   Supply and Ground Requirements

The supply (V_{DD}) and ground (GND) sources require pull-up and pull-down networks respectively to provide a conducting path for charging and discharging the output node. Power and ground terminals are presented with keywords **supply1** and **supply0**.

Example:
A CMOS inverter is as shown in Figure 7.2.
Module coms_inv(OUT, IN);

```
supply1 V_DD;              //logic 1 connected to VDD
supply0 GND;               //logic 0 connected to GND
pmos p(OUT, V_DD, IN);     //instantiate PMOS switch
nmos n(OUT, GND, IN);      //instantiate NMOS switch
endmodule
```

7.5 Logic Implementation Using Switches

A few examples are given in detail here for the implementation of any function using the switch-level model in Verilog HDL.

NAND Gate: For the design of 2-input NAND gates, two PMOS and two NMOS transistors are required for pull-up and pull-down networks, respectively. CMOS implementation always provides complemented output. For NAND operation, NMOS are connected in series while PMOS are connected in parallel.

With the help of Table 7.1, a NAND gate can be designed at the transistor level. Figure 7.6 presents the implantation of a NAND gate using MOS switches.

Table 7.1 Truth table of a NAND gate.

A	B	$F = \overline{A.B}$
0	0	1
0	1	1
1	0	1
1	1	0

// Verilog Program for NAND using switches

```
module
NAND_Switch(F,A,B);
input A,B;
output F;
wire w1;
supply1 Vdd;
supply0 gnd;
pmos P1(F,Vdd,A);
pmos P2(F,Vdd,B);
nmos N1(F,W1,A);
nmos N2(W1,gnd,B);
endmodule
```

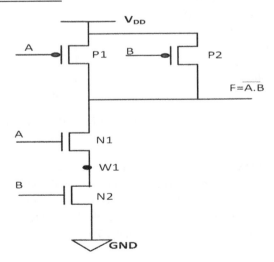

Figure 7.6 NAND gate implantation at transistor level.

The AND gate can be designed by adding one inverter stage at the output of the NAND gate. So, overall, six transistors including three NMOS and three PMOS, are required to design AND gates using switches. With the help of Table 7.2, the AND gate can be implemented at the transistor level as shown in Figure 7.7

Table 7.2 Truth table of an AND gate.

A	B	F = A.B
0	0	0
0	1	0
1	0	0
1	1	1

// Verilog Program for AND using switches

```
module
AND_Switch(F,A,B);
input A,B;
output F;
wire w1, F1;
supply1 Vdd;
supply0 gnd;
pmos P1(F1,Vdd,A);
pmos P2(F1,Vdd,B);
nmos N1(F1,W1,A);
nmos N2(W1,gnd,B);
pmos P3(F, VDD, F1);
nmos P4(F, GND, F1);
endmodule
```

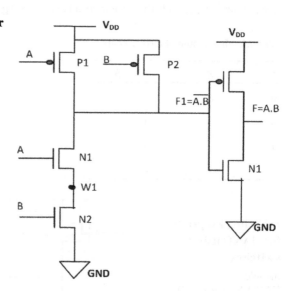

Figure 7.7 AND gate using MOS switches.

NOR Gate: For the design of 2-input NOR gates, two PMOS and two NMOS transistors are required for pull-up and pull-down network, respectively. For a NOR operation, NMOS are connected in parallel while PMOS are connected in series. With the help of Table 7.3, the NOR gate is implemented at the level of the transistor as shown in Figure 7.8.

Table 7.3 Truth table of a NOR gate.

A	B	$F = \overline{A+B}$
0	0	1
0	1	0
1	0	0
1	1	0

// Verilog Program for NOR using switches

```
module swichnor(out,a,b);
output out;
input a, b;
//internal wires
wire c;
supply1 VDD; //power (1) is connected
to VDD
supply0 GND; //ground (0) is connected
to GND
pmos (out, VDD, A);
pmos (out, VDD, B);
nmos (out, GND, A);
nmos (out, W1, B);
endmodule
```

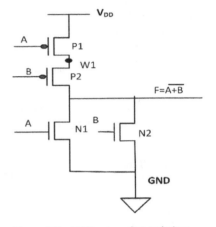

Figure 7.8 NOR gate using switches.

With the help of Table 7.4, an OR gate can be designed by adding one inverter stage at the output of a NOR gate.

Table 7.4 Truth table of an OR gate.

A	B	$F = A+B$
0	0	0
0	1	1
1	0	1
1	1	1

// **Verilog Program for OR using switches**

```
module
OR_Switch(F,A,B);
input A,B;
output F;
wire w1, F1;
supply1 VDD;
supply0 GND;
pmos P1(W1,VDD,A);
pmos P2(F1,W1,B);
nmos N1(F1,GND,A);
nmos N2(F1,GND,B);
pmos P3(F, VDD, F1);
nmos P4(F, GND, F1);
endmodule
```

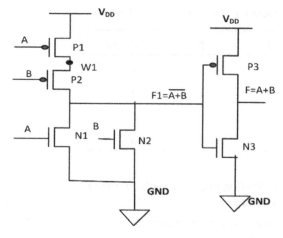

Figure 7.9 OR gate using switches.

With the help of Table 7.5, **XOR gate**, an XOR gate can be implemented using switches with AND, OR and NOT operations as shown in Figure 7.10.

// **Switch XOR**

```
module switchxor(f,a,b);
input a,b;
output f;
wire w1,w2,w3,w4,w5;
supply1 vdd;
supply0 gnd;
nmos (w3,gnd,w2);
nmos (f,w3,w1);
nmos (w4,gnd,b);
nmos (f,w4,a);
pmos (f,w5,a);
pmos (f,w5,b);
pmos (w5,vdd,w1);
pmos (w5,vdd,w2);
inv n1(w1,a);
inv n2(w2,b);
endmodule
module inv(w1,a);
input a;
output w1;
supply1 vdd;
supply0 gnd;
nmos (w1,gnd,a);
pmos (w1,vdd,a);
endmodule
```

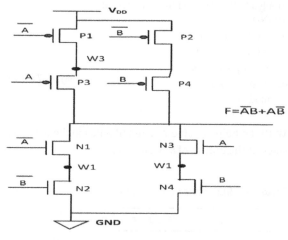

Figure 7.10 XOR gate using switch.

Table 7.5 Truth table of an XOR gate.

A	B	$F = \bar{A}B + A\bar{B}$
0	0	0
0	1	1
1	0	1
1	1	0

7.5.1 Digital Design with a Transmission Gate

CMOS transmission gates are used to build logic function at the level of transistors. With transmission gates, the logic implementations are carried out by the appropriate setting of control signals.

7.6 Implementation with Bidirectional Switches

Transmission gates are examples of bidirectional switches. They can be resistive or non-resistive. Resistive switches reduce the signal strength due to their high impedances. CMOS is one of the examples of a bidirectional switch that can be used for the implementation of any Boolean function with a suitable p- and n-control signal.

7.6.1 Multiplexer Using Switches

Multiplexers are the data selector that selects a particular input value according to the value of select lines. In the switch-level model, it can be implemented using either MOS or CMOS switches. Table 7.6 presents the truth table of a 2×1 multiplexer.

Figure 7.11 shows a 2×1 multiplexer block. Figure 7.12 presents a 2×1 multiplexer using CMOS switches.

Table 7.6 Truth table of an OR gate.

Select(S)	*Out*
0	I0
1	I1

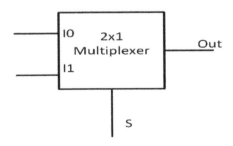

Figure 7.11 2×1 Multiplexer block.

// 2x1 Multiplexer using CMOS switch

```
module mux21_switch(out, s, i0, i1);
output out;
input s,i0,i1;
wire s1;
snot u(s1,s);
cmos (out,i0,s1,s);
cmos (out,i1,s,s1);
endmodule
module snot(s1, s);
output sbar;
input s;
supply1 vdd;
supply0 gnd;
nmos (s1,gnd,s);
pmos (s1,vdd,s);
endmodule
```

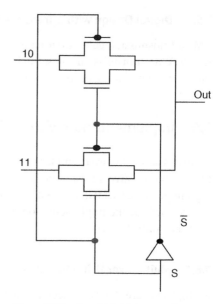

Figure 7.12 2 × 1 Multiplexer using a CMOS switch.

A 4 × 1-multiplexer block has been shown in Figure 7.13. Using the truth table 4 × 1 multiplexer, a switch-level module has been designed using CMOS switches at the level of the switch-level model. Two select lines S0 and S1 and I0, I1, I2, I3 are considered as data lines. To select a particular data line, a suitable value of select lines are required. So, the control signal of CMOS switches will be connected to achieve the desired output function. Figure 7.13 describes a 4 × 1-multiplexer block with 4-input, 2-select, and 1-output lines. With the help of Table 7.7, a 4 × 1 multiplexer is implemented in Figure 7.14 at the transistor level.

Table 7.7 Truth table of a 4 × 1 multiplexer.

S1	S0	Y
0	0	I0
0	1	I1
1	0	I2
1	1	I3

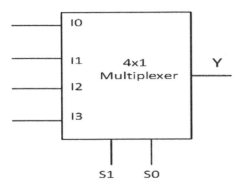

Figure 7.13 4 × 1-multiplexer block.

// 4x1 Multiplexer using CMOS switch
```
module mux41_switch (Y, S0,
S1, I0, I1, I2, I3);
output out;
input I0, I1, I2, I3, S0, S1;
wire w0,w1;
NOT_switch n1(w0, S0);
NOT_switch n2(w1, S1);
cmos c1(Y, I0,w1, s0);
cmos c2(Y, I1, w1, w0);
cmos c3(Y,I2,S1,S0);
cmos c4(Y,I3,S1,w0);
endmodule
module NOT_switch(S_bar, s);
output y;
input s_bar;
supply1 vdd;
supply0 gnd;
nmos (s_bar,GND,s);
pmos (s_bar,VDD,s);
endmodule
```

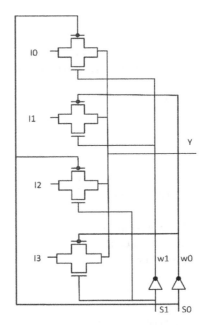

Figure 7.14 4 × 1-multiplexer using switches.

Example:

Different switches can also be used in the same module. As in the implementation of a full adder, we can use a 4 × 1 multiplexer implemented with nmos, pmos, and cmos switches. In this, the top-level full-adder block (Figure 7.15) is divided into a sub-block of two 4 × 1 multiplexers with the help of switches as shown in Figure 7.14.

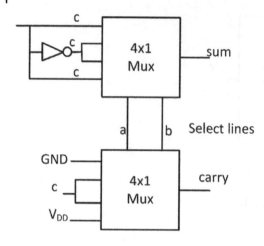

Figure 7.15 1-bit full-adder implementation using a 4×1 multiplexer.

```
//1-bit Full adder block using 4×1 multiplexer
module fulladder_Mux41_cmos(a,b,cin,sum,carry);
input a,b,cin;
output sum,carry;
supply0 GND;
supply1 VDD;
wire w1;
not n1(w1,cin);
mux41 x1(a,b,cin,w1,w1,cin,sum);
mux41 x2(a,b,GND,cin,cin, VDD,carry);
endmodule
//4x1 Multiplexer block using switches.
module mux41(s0,s1,i0,i1,i2,i3,y);
input s0,s1,i0,i1,i2,i3;
output y;
supply0 GND;
supply1 VDD;
wire w1,w2,w3,w4;
nmos n1(w1, VDD,s0);
pmos p1(w1,GND,s0);
nmos n2(w2, VDD,s1);
pmos p2(w2,GND,s1);
cmos c1(w3,i0,s0,w1);
cmos c2(w3,i1,w1,s0);
cmos c3(w4,i2,s0,w1);
cmos c4(w4,i3,w1,s0);
cmos c5(y,w3,s1,w2);
```

```
cmos c6(y,w4,w2,s1);
endmodule
//Verilog test stimulus for 1-bit Full adder
module fulladder_cmos_test;
reg [7
:0]a,b,cin;
wire sum,carry;
cmosfulladd f1(a,b,cin,sum,carry);
initial begin
a = 0;b = 0;cin = 0;
#3 a = 0;b = 0;cin = 1;
#3 a = 0;b = 1;cin = 0;
#3 a = 0;b = 1;cin = 1;
#3 a = 1;b = 0;cin = 0;
#3 a = 1;b = 0;cin = 1;
#3 a = 1;b = 1;cin = 0;
#3 a = 1;b = 1;cin = 1;
#3 $stop;
end endmodule
```

7.7 Verilog Switch-Level Description with Structural-Level Modeling

Verilog HDL supports the structural-level model with modules at the switch level. The implementation of the complex digital circuit can easily be implemented at the structural level including the switch level in the submodule. Suppose the designer wants to implement a 4-bit ripple carry full adder, then the main module must be subdivided into the submodules of the 1-bit full adder and then the 1-bit full adder can be elaborated with a logic gate that can easily be implemented with nmos and pmos switches. This indicates that, with the help of a structural-level model, a complex top module can be simplified in smaller sub-blocks and a further declaration sub-block at a root level of models like switches by exploiting the top-down design methodology.

7.8 Delay Model with Switches

Delays can be specified for signals that pass through these switch-level elements such as MOS, CMOS, and bidirectional switches. Delay specification is optional and it can be applied immediately after the Verilog primitive or keyword for the switch. Just as in the gate-level model, rise-fall delay, rise-fall-turnoff delay or min-typical-max delays can also be specified in the switch-level model. Rise and fall specify the delay associated with output transition from 0-1 and 1-0 zero transitions. In some designs, the minimum and maximum delays are specified or provided to the designer.

Example:

pmos p1(Drain, Source, gate);
pmos #(1) p1(Drain, Source, gate);
nmos #(1, 2) p2(Drain, Source, gate);   //Delay specified by rise and fall time
nmos #(1, **3, 2)** p2(Drain, Source, gate);   //Delay specified by rise, fall and turnoff
                                                       time

cmos #(5) c2(out, in, ngate, pgate);

Discussion in this chapter focuses on the design complexity that increases at the level of the transistor and is also difficult to recall the interconnections of transistors in a complex circuit. However, the understanding of such design and implementation is important for beginners to understand the actual logic transition happening inside the logic circuit. It also provides greater insights on the charging and discharging of the output node depending on input combinations. So, the switch-level implementation using the Verilog module is equally important for the strong foundation of learners and designers in the area of VLSI design.

Index terms: CMOS (NMOS/PMOS) design, switch-level modeling, transmission gate-based design.

Review Questions

Q1 Write a Verilog program for three input NAND/NOR gates using switches.

Q2 Implement a half subtractor using NMOS and PMOS switches and write Verilog code for the same.

Q3 Implement half adder using NMOS and PMOS switches and write Verilog code for the same.

Q4 Implement the compliment of $F = (a + b).(c + d)$ using NMOS and PMOS switches and write Verilog code using the switch-evel model.

Q5 Implement the compliment $F = a.b + b.c + a.c$ using NMOS and PMOS switches and write Verilog code using the switch-level model.

Q6 Implement a 2-to-4 line decoder using CMOS switches and write a Verilog program and test stimulus for the same.

Q7 What do you mean by resistive switches? Give a suitable example for the switch-level model.

Multiple Choice Questions

Q1 The minimum number of transistors required for XNOR gate implementations using CMOS technology are:
  A 7
  B 8
  C 9
  D 10

Q2 The correct syntax for tranif1 and transif0 are:
  A tranif1 (prt1, prt2, ctrl); tranif0(prt1, prt2, *ctrl*);
  B tranif1 (prt1, prt2, ctrl); tranif0(prt1, prt2,ctrl);
  C tranif1 (prt1, prt2, *ctrl*); tranif0(prt1, prt2, ctrl);
  D all of the above

Q3 Choose the correct option for the highest and lowest level of abstraction in Verilog HDL
  A dataflow model, gate-level model
  B dataflow model, switch-level model
  C switch-level model, behavioral model
  D behavioral model, switch-level model

Q4 Which is incorrect for the switch-level model?
  A Lowest level of abstraction in Verilog HDL
  B Delay can't be specified in switch level
  C Verilog supports bidirectional switches
  D Interconnections are required at the transistor level

Q5 Which is an incorrect example of bidirectional switches?
  A tran
  B CMOS
  C tranif1
  D notif1

Q6 Which is an incorrect option for resistive switches?
A High impedance
B Low impedance
C rtran
D rnmos

References

[1] Man, M. and Ciletti, M.D. (2013). *Digital Design: With an Introduction to the Verilog HDL*, 5e. ed. Upper Saddle River, NJ: Pearson.

[2] Palnitkar, S. (2001). *Verilog HDL*. Upper Saddle River, NJ: Pearson.

8

Advance Verilog Topics

8.1 Delay Modeling and Programming

In actual hardware, different delays exist and broadly, delays can be divided into two types i.e., gate delays and wire or interconnection delays. During the design of any circuit, delays must be considered in the programming to match with actual hardware timings. With advancement of technology, timing constraints become crucial and hence need to be taken care of while programming. Verilog also has the facility to introduce delay while designing the circuits [1–2].

8.1.1 Delay Modeling

Three different types of delay modeling are available in Verilog:
(i)   Distributed-delay modeling
(ii) Lumped-delay modeling
(iii) Pin-to-Pin-delay modeling or Path-delay modeling.

8.1.2 Distributed-Delay Model

This type of delay is specified with each element of the circuit. In the case of combinational circuits, each gate has its propagation delay and is specified within the statement of all gates used in the gate-level modeling style of Verilog.

As an example in the following circuit diagram shown in Figure 8.1, the distributed delay is indicated inside the image of all the gates. The Verilog code for the same is shown in Program P8.1.

Digital VLSI Design and Simulation with Verilog, First Edition. Suman Lata Tripathi, Sobhit Saxena, Sanjeet Kumar Sinha, and Govind Singh Patel.
© 2022 John Wiley & Sons Ltd. Published 2022 by John Wiley & Sons Ltd.

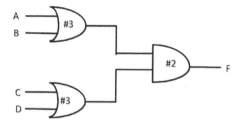

Figure 8.1 A simple combinational circuit indicating distributed delay.

```
//Program P8.1
module distributed_delay (F,A,B,C,D);
output F;
input A,B,C,D;
wire w1,w2;
or #3 x1(w1,A,B);
or #3 x2(w2,C,D);
and #2 x3(F,w1,w2);
endmodule
```

The distributed delay can also be specified in the dataflow modeling style within the assignment statement as shown in Program P8.2 which is a re-written program of Program P8.1 in dataflow style for the same circuit shown in Figure 8.1.

```
//Program P8.2
module distributed_delay (F,A,B,C,D);
output F;
input A,B,C,D;
wire w1,w2;
assign #3 w1 = A&B;
assign #3 w2 = C&D;
assign #2 F = w1&w2;
endmodule
```

8.1.3 Lumped-Delay Model

Lumped delay is the overall delay associated with a particular circuit or module. As shown in Figure 8.2, only one delay is specified in the gate whose output is the final output of the circuit, which is calculated based on the overall delay from input to output (see Figure 8.1). Lumped delay is normally specified on the output gate. The Verilog code for the same is specified in Program P8.3.

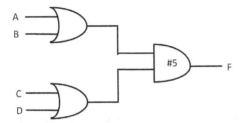

Figure 8.2 A simple combinational circuit indicating lumped delay.

```
//Program P8.3
module lumped_delay (F,A,B,C,D);
output F;
input A,B,C,D;
wire w1,w2;
assign w1 = A&B;
assign w2 = C&D;
assign #5 F = w1&w2;
endmodule
```

If we compare Program P8.2 and Program P8.3, it can easily be seen that the delay specified for the AND gate is the sum of delays associated with the OR gate and AND gate.

8.1.4 Pin-to-Pin-Delay Model

Delay can also be specified in the pattern of delays from individual input to output. This type of delay specification is known as pin-to-pin delay or path delay. Figure 8.3 represents the possible paths from individual inputs to the output of the circuit shown in Figure 8.2.

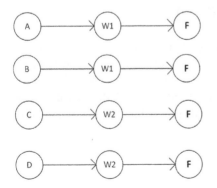

Figure 8.3 Possible path from individual input to output.

Let us assume that the associated delay with each path is as follows:

path A-w1-F, delay = 5

path B-w1-F, delay = 6

path C-w2-F, delay = 7

path D-w2-F, delay = 8

Pin-to-pin delay or path delay is specified in the Verilog module using specific blocks. The specify block in any module can be introduced by using keywords **specify** and **endspecify.**

The statements for pin-to-pin delays or path delays can be written within these keywords. Program P8.4 shows the format of **specify** block.

```
//Program P8.4
module path_delay (F,A,B,C,D);
output F;
input A,B,C,D;
wire w1,w2;
specify
(A = > F) = 5;
(B = > F) = 6;
(C = > F) = 7;
(D = > F) = 8;
endspecify
assign w1 = A&B;
assign w2 = C&D;
assign #5 F = w1&w2;
endmodule
```

8.2 User-Defined Primitive (UDP)

Verilog has a number of built-in primitives that can be illustrated in modules viz. gate type (AND, NAND, OR, NOR, XOR, XNOR, etc.) and switch type (NMOS, PMOS, CMOS, etc.). Verilog enables the user to create their own primitives called user-defined primitives (UDP). In UDP, the user describes any logic with the help of a truth table. The syntax for the same is shown in Program P8.5.

//Program P8.5

Primitive name_of_UDP (output, input1, input2);

Declaration_of_ports

reg output;//optional

initial

output = some_initial_value;//optional

table

truthtable

endtable

endprimitive

Any UDP should begin with the keyword **primitive** followed by its name (selected by the user). Only one output followed by the number of inputs should be provided in the bracket. The declaration of ports as output or input need to be declared in the port-declaration section. Defining output port as **reg** and its initialization with some value is optional. This needs to be done in the case of sequential UDPs. The programming architecture of UDPs is defined in the form of a truth table from which the value of output can be determined for all possible combinations of the inputs. This starts with a keyword **table** and ends with the keyword **endtable**. The format of the table will be discussed in the upcoming section. UDPs can be divided into two types: combinational and sequential UDP.

8.2.1 Combinational UDPs

Combinational UDPs are those UDPs in which the output is determined by a logical combination of the inputs as in any combinational circuit. The only difference is that the UDP has only one output but the combinational circuit can have more than one.

8.2.1.1 Truth-Table Format

To understand the format of writing a truth table between keywords **table** and **endtable**, let's start by creating a UDP of a 2-input AND gate and give it the name my_and. The UDP is shown in Program P8.6.

//Program P8.6

primitive my_and (f, a, b);

output f;

input a, b;

table

//a b: f

0 0: 0;

0 1: 0;

1 0: 0;

1 1:1;

endtable

endprimitive

In the above UDP of an AND gate, it can be seen that the line next to the keyword **table** is a comment (indicated by//) which means the name of the ports a, b, and f are not allowed to write inside the table. A few important points that need to be remembered while writing the truth table are:

i) The sequence in which the values are written should be in the same sequence in which the input ports are declared.

ii) Values assigned to input ports are separated by a space.

iii) The value assigned to output should be separated from input values by a semicolon ":".

Also, note that if the output value is not defined in the truth table for any input combination and that input combination occurred during simulation, **x** will be assigned to the output.

8.2.1.2 Values Assigned to Input Ports

In section 8.2.1, we have learned how to write a truth table in a UDP. Until now, we have considered values "0" and "1" only for the input. We are now going to consider other values "x" and "z" that are allowed in Verilog coding. Let's rewrite the AND gate program again named as my_and2.

//Program P8.7

primitive my_and2(f, a, b);

output f;

input a, b;

table

//a b: f

0 0: 0;

0 1: 0;

1 0: 0;

1 1:1;

0 x: 0;

x 0: 0;

endtable

endprimitive

We can see in Program 8.7, that some entries include "x" at the input, but the output for the same is "0". This is because we know from the property of an AND gate that if any input is "0" output must be "0". For other combinations where one input is "1" and other input is "x" or both inputs are "x," the output will be "x," that's why this has not been included in the table, as we know that, for the input combinations not available in the table, output will be by-default "x." Any "z" value will be considered as "x."

A shorthand notation "?" is also available which we can use in UDP to make a truth table to reduce the size of the table. An example of a 2 × 1 multiplexer is shown in Program P8.4 using "?." Here, the "?" represents all possible values "0","1" and "x." So, it reduces four different entries in the table to just one.

```
//Program P8.8
primitive mux2x1 (f, a, b, s);
output f;
input a, b, s;
table
//a b s: y
0 ? 0: 0;//"?" can be "0" or "1" or "x"
1 ? 0: 1;//"?" can be "0" or "1" or "x"
? 0 1: 0;//"?" can be "0" or "1" or "x"
? 1 1:1;//"?" can be "0" or "1" or "x"
endtable
endprimitive
```

The functionality of a multiplexer is to connect one input (out of many inputs) to the output, one at a time. The input is selected according to the value fed to the select line. So, if s = "0" whatever the value on input, a goes to the output independent of the value at b. Thus "?" is best suited value for b in the table where s = "0." Similarly "?" is the best-suited value for a in the table where s = "1" as indicated in Program P8.8.

8.2.1.3   Instantiation of a Combinational UDP

The UDP can be instantiated in any Verilog module like a keyword. It is not compulsory to give the instance any name as in the case of a keyword. Let's write a module for a half-adder circuit using the UDP created in Program P8.7.

```
//Program P8.9
module half_adder (a,b,sum,carry);
input a,b;
output sum,carry;
```

xor (sum,a,b);

my_and2 (carry, a,b);

endmodule

8.2.2 Sequential UDPs

Sequential UDPs are those UDPs in which the next output is decided by the combination logic of the current inputs and current output. Sequential UDPs are different from combinational UDPs in the following terms:

(i)   the output port needs to be declared as **reg**.
(ii) the output needs to be initialized by some value, either "0" or "1."
(iii) the values of inputs, current output, and next output indicated in the table are separated by a semicolon.

The sequential UDPs can be further divided into two types: Level sensitive and edge sensitive.

8.2.2.1   Level-sensitive Sequential UDPs

Level-sensitive sequential UDPs are those UDPs in which the next output depends on the level of the clock. If the level of the clock is "1," the next output is computed according to current inputs and output but if the clock is "0," the next output retains the value of the current output. As an example, the level-sensitive sequential UDP of a D flip-flop (F/F) is presented below in Program P8.10.

```
//Program P8.10
primitive dff (q, d, clock, rst);
output q;
input d, clock, rst;
reg q;
initial
q = 0;
table
//d clock rst: current q: next q
? ? 1: ?: 0;//reset condition
0 1 0: ?: 0;//clk level high
1 1 0: ?: 1;//clk level high
? 0 0: ?: -;//clk level low
endtable
endprimitive
```

D-F/F UDP, named as dff, has one output port q and three input ports d, clock, rst. Here, we can see that output port q needs to be declared as **reg** as it has to store the current value of the output (current q). The output q is also initialized with the value "0" which becomes the value of current q. If the reset (rst) is "1," then the value of the next output will be "0" independent of the values of d, clock (clock), and current output. This is called the reset condition. If rst is "0" then its behavior depends on the level of the clk. If clock = "1" whatever on d goes to the next output, otherwise the next output retains the value of the current output. "-" symbol is used to indicate retention of the current output as the next output.

8.2.2.2 Edge-sensitive Sequential UDPs

Edge sensitive means the next output depends on the edge of the clock pulse (either + ve edge 0→1 or –ve edge 1→0). These are called + ve edge triggered or –ve edge triggered depending on whether the next output appears on the + ve or –ve edge of the clock, respectively.

As an example of a –ve edge triggered D-F/F, the UDP is shown below in Program P8.11.

```
//Program P8.11
primitive EDff (q, d, clk, rst);
output q;
input d, clk, rst;
reg q;
initial q = 0;
table
//d clk rst: current q: next q
? ? 1: ?: 0;         //reset condition
? (0?) 0: ?: -;      //ignore + ve transition of the clk
? (x1) 0: ?: -;      //ignore unknown + ve transition of the clk
? (1x) 0: ?: -;      //ignore unknown –ve transition of the clk
1 (10) 0: ?: 1;      //next output appears at –ve edge of the clk
0 (10) 0: ?: 0;      //next output appears at –ve edge of the clk
(??) 0 0: ?: 0;      //ignore change in d when clk is stable
endtable
endprimitive
```

The transition of any value can be represented as (v1v2) in Table 8.1, where v1 and v2 are the two different values out of "0,"1," and "z." EDff represents the

UDP of a -ve edge triggered D-F/F in which all possible transitions in the clk are being considered in the table. When the clk is stable and transition occurs at d, no change in the output takes place and is hence ignored. It is also important to note that transition at two different ports is not allowed at the same time.

8.2.3 Shorthands in UDP

The different shorthands that can be used in creating the UDP are shown in Table 8.1. So, wherever we have used the shorthand symbol, the meaning can be obtained from this table (if not explained in the text, try to use these symbols in creating different UDPs).

Table 8.1

Shorthand symbol	Meaning
?	0 or 1 or x
-	Retention of current state
*	(??)
B	0 or 1
R	(01)
F	(10)
P	(01) or (0x) or (x1)
N	(10) or (1x) or (x0)

8.3 Task and Function

While undertaking a design project in Verilog, a user may have to implement the same activity/steps at different places. It is therefore required to save each particular functionality at a particular location and invoke or call the same functionality whenever required in the main program instead of repeating the same code every time. These types of programs/codes are generally known as subroutines and are facilitated in almost all languages. In Verilog, task and function are available to fulfill the requirements of subroutines.

8.3.1 Difference between Task and Function

A task is defined as a set of instructions/statements for performing a particular functionality and that will be used in the design repeatedly. For example, a particular task includes shifting the contents from one memory location to another, performing some arithmetical operations and saving the result at a different memory location.

A function is normally used to evaluate an expression/formula which gives a single output for a particular set of inputs. For example, a function can be created for an equation $4 \times {}^2 + 3y^2 + xy$ which needs to be evaluated many times for the different values of x and y. That is why the function returns only one single value when invoked.

The differences between task and function are summarized in Table 8.2. Understanding the differences mentioned in Table 8.2 will be more effective when we consider the examples given in the upcoming sections.

Table 8.2 Differences between task and function.

Task	Function
Task can enable other task or function.	A function cannot enable any task but can enable another function.
Simulation time for the execution of task can be non-zero.	Simulation time for the execution of a function is always zero.
Tasks may contain event, timing-control, or delay statements.	Functions must not contain any event, timing-control, or delay statements.
Tasks may have no output, input or inout, or many of them.	At least one input is mandatory in function. It can have more than one input. Output or inout are not allowed.
No return value is there in a task but multiple values can be passed.	Functions always return one single value.

Some similarities between task and function are:

(i) Both function and task are defined in a module and belong to that particular module.

(ii) Wires are not used in any task or function. They are constructed only from behavioral statements.

(iii) *Initial* and *always* statements are not used in task and function.

(iv) Invoking of task and function is done inside the *initial* and *always* block or via another task or function.

8.3.2 Syntax of Task and Function Declaration

//Program P8.12 Task_Declaration_Syntax

task <task_name>;

```
< input-declaration>;
< output-declaration>;
< inout-declaration>;
< parameter-declaration>;
< reg-declaration>;
< time-declaration>;
< integer-declaration>;
< real-declaration>;
< event-declaration>;
begin
<statement_or_null>;
end
endtask
```

Declaration of the task starts with the keyword **task** and ends with keyword **endtask**. Task declaration is described in detail via Program P8.12. The name of the task is declared just after the keyword **task** followed by a semicolon (;). Input, output, and inout are declared to pass the values to and from the task. The rest are declared as per the requirement in the task statements. All declarations in any task are optional including input, output, and inout. A sequence of statements describing the task are declared between begin and end.

```
//Program P8.13 Function_Declaration_Syntax
function <range_or_type> <function-name>;
< input-declaration>;
< parameter-declaration>;
< reg-declaration>;
< time-declaration>;
< integer-declaration>;
< real-declaration>;
<statement including function-name>
endfunction
```

Declaration of function starts with keyword **function** and ends with keyword **endfunction**. Function declaration is described in detail via Program P8.13. The name of the function is very important because the return value is assigned to the function-name. That is why range or type can also be declared with the name of the function. Only input declaration is there and there is no output or inout. The rest are declared as per the requirement in the function statements. The function statement must include "function-name=" followed by the expression of input combinations. At least one input is mandatory in any function. A sequence of statements describing the function are declared between begin and end.

8.3.3   Invoking Task and Function

Declared task and function can be invoked in any module using the format given in Programs P8.14 and P8.15, respectively.

//Program P8.14 Task invocation Syntax

task_name (< outputs_from_the_task>, < input_to_the_task>);

//Program P8.15 Function Invocation Syntax

name_of_function (<inputs_to_the_function>);

8.3.4   Examples of Task Declaration and Invocation

To understand the declaration and invocation of task examples of task bitwise_ oper is shown in Program P8.16.

//Program P8.16

module operation(ab-and, ab-or, ab-xor,a,b);

input [15:0] a, b;

output [15:0] ab-and, ab-or, ab-xor;

reg [15:0] ab-and, ab-or, ab-xor;

always @(a or b)

begin

bitwise_oper(ab-and, ab-or, ab-xor, a, b);//Task Invocation

end

task bitwise_oper;//Task declaration

output [15:0] ab-and, ab-or, ab-xor;

input [15:0] a, b;

begin

#10 ab_and = a & b;

ab_or = a | b;

ab_xor = a ^ b;

end

endtask

endmodule

8.3.5 Examples of Function Declaration and Invocation

To understand the declaration and invocation of task and function, an example of function check_parity is shown in Program P8.17.

```
//Program P8.17
module parity(parity,address);
input [31:0] address;
output parity;
reg parity;
always @(address)
begin
parity = check_parity(address); //function invocation
$display("Parity calculated = %b", check_parity(address)); //function
invocation
end
function check_parity; //function declaration
input [31:0] addr;
begin
check_parity = ^address;
end
endfunction
endmodule
```

Review Questions

Q1 Delay modeling is important in Verilog HDL, why?

Q2 UDPs are important, why? Give any two advantages of creating UDPs in Verilog HDL.

Q3 How does the UDP for sequential circuits differ from the combinational circuit? Explain the answer with a suitable example.

Q4 Write a Verilog program of full adder using multiplexer as UDP.

Q5 Write a Verilog program of full subtractor using UDP of NOT, AND, OR and Ex-OR gate.

Multiple Choice Questions

Q1 The correct statement for lumped delay is:
- **A** the delay is specified per element basis
- **B** the delay is specified per module
- **C** the delay is specified for specific path
- **D** none of the above

Q2 Path delays are assigned in Verilog within the keywords:
- **A** table, endtable
- **B** fork, join
- **C** task, endtask
- **D** specify, endspecify

Q3 The incorrect statement for combinational UDPs:
- **A** It can be used to create a library of user-defined primitives
- **B** The output in combinational UDPs must be declared on reg
- **C** The table can be inserted in combination UDPs with keyword *table, endtable*
- **D** UDPs can be declared as inside the keyword *primitive, endprimitive*

Q4 The correct syntax of writing a truth table in sequential UDPs is:
- **A** in1 in2: Q;
- **B** in1 in2 clk: Q:Q+;
- **C** in1 in2 (10);Q;Q+;
- **D** in1, in2, clk:Q:Q+;

Q5 The meaning of (1x) related to clock transition is:
- **A** denotes a transition from *0* to 0, 1, or X. Potential positive-edge transition.
- **B** denotes a transition from *x* to 0, 1. Potential negative-edge transition.
- **C** denotes any transition in signal value *0*, 1, or X to 0,1, or *X*.
- **D** denotes a transition from logic 1 to unknown X state.

References

[1] Palnitkar, S. (1996). *Verilog HDL*. Upper Saddle River, NJ: Sunsoft Press.
[2] Bala Tripura Sundari, B. and Padmanabhan, T.R. (2003). *Design through Verilog HDL*. Hoboken, NJ: Wiley.

9

Programmable and Reconfigurable Devices

9.1 Logic Synthesis

The design of a complex function is mainly performed through a higher level of abstraction such as behavioral or dataflow models. In logic synthesis, the designer works on getting the transistor level net list from the register transfer level (RTL) or a higher level of modules. The advantage of these higher levels is that a designer can easily solve and implement any digital circuit at algorithm or truth table with the knowledge of certain design specifications. The process of logic synthesis helps in achieving an optimum gate level netlist or logic implementation through higher-level Verilog models without any concern for internal connections, wires, and the number of components in comparison to the lower-level modules of gate-level or switch-level models. Logic synthesis plays an important role in complex digital circuit design in obtaining optimum hardware elements and interconnects. There are several ways to implement a particular digital function within the specified design constraints. So, getting an optimum design is a highly automated process that bridges the gap between high-level design synthesis and the physical-layout processes. Figure 9.1 shows a general workflow in logic synthesis. Logic synthesis includes standard cell libraries.

9.1.1 Technology Mapping

Technology mapping is an important aspect that sets a coherence between initial logic descriptions and the target technology. Prior to technology mapping, design descriptions were independent of any target technology. Therefore,

Digital VLSI Design and Simulation with Verilog, First Edition. Suman Lata Tripathi, Sobhit Saxena, Sanjeet Kumar Sinha, and Govind Singh Patel.
© 2022 John Wiley & Sons Ltd. Published 2022 by John Wiley & Sons Ltd.

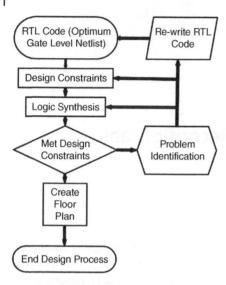

Figure 9.1 VLSI design flow at RTL level.

technology mapping is one of the design steps of logic synthesis where internal gate-level representation is carried out using a cell in the standard technology library.

9.1.2 Technology Libraries

As per the requirement of a particular design functionality, the designer can use or create a specific technology library that may be able to provide all logic functionality and other input/output requirements. The technology library constitutes different library elements such as primitives and macros. In an Xilinx library, certain primitives can be the basic circuit elements such as logic gates (AND, OR, NOT, etc.). The primitive has a certain symbol with a unique name and description. Macros constitute a few elements such as macros and primitives that are picked from the library itself.

9.2 Introduction of a Programmable Logic Device

In general logic devices, the hardware elements and connections are fixed to perform certain functions, for example, basic gates or any fixed type of combinational- and sequential-circuit design. Programmable logic devices (PLDs) came into the picture for complex logic design where it was difficult to make onboard connections for functional design and verification processes. PLDs can be programmed after manufacture to obtain different logic functions. With the help of programmable devices, we can also develop prototype hardware for certain complex logic function implementation before fabrication. Also, the programmable devices can easily deploy for different applications as per requirement such as ready-to-use design board facilitating logic elements, input/output and programmable interconnecting wires. The programmable devices are categorized depending on their work, the number of hardware elements and design complexity. A few of them are listed in the next section.

9.2.1 PROM, PAL, and PLA

Programmable ROM (PROM): This is a type of non-volatile read-only-memory (ROM) that can be programmed to write the content permanently by burning out the programmable switches (fuse). The PROM is a simple block of fixed AND-plane followed by a programmable OR-plane used to obtain any desired logic function. Let us take one example of 3-input PROM block as depicted in Figure 9.2.

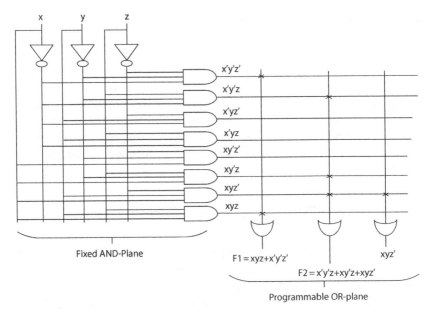

Figure 9.2 Example of function implementation with PROM.

Programmable Array Logic (PAL): PAL block is a comparatively simpler PLD that has two levels in the form of programmable AND-plane connected to a fixed OR-plane. A desired logic function is implemented by applying suitable values to programmable switches connected to an AND-plane. Figure 9.3 shows an example of 3-variable function implementation with PAL.

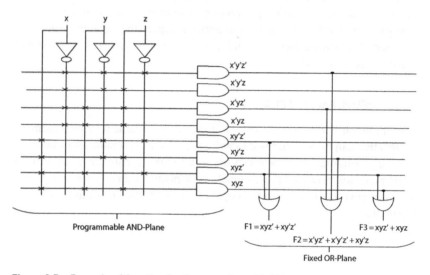

Figure 9.3 Example of function implementation with PAL.

Programmable Logic Array (PLA): The PLA block is a comparatively simpler PLD that has two logic planes in the form of programmable AND-plane and programmable OR-plane. In PLA, a designer can program both AND-plane and OR-plane for different logic functionality with suitable input values to programmable switches. PLA structures can also be a part of full-custom chips. Figure 9.4 shows an example of 3-variable function implementation PLA.

9.2.2 SPLD and CPLD

Simple PLD (SPLD): Simple PLD is similar to the PLA or PAL blocks. SPLDs (PLA, PAL) are usually available with a limited number of input and output pins (I/O pins <32 and so on) and product terms.

Complex PLD (CPLD): As the name indicates, CPLDs are with a greater number of input/output pins or other blocks such as multiple SPLD blocks on a single chip. Alternative names are sometimes adopted for the chip as enhanced PLD (EPLD), Super PAL, and Mega PAL, etc.

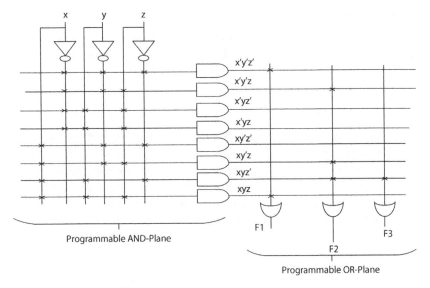

Figure 9.4   Example of function implementation PLA.

The following are the major blocks of CPLD:

- It has a PAL-like block.
- Programmable interconnecting switches.
- I/O block that is connected to PAL through programmable switches.

Here, the PAL block shown in Figure 9.5 consists of the following major components:

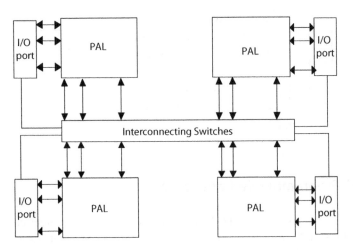

Figure 9.5   CPLD block diagram.

i) Programmable AND Plane

Programmable AND plane is similar to $n:2^n$ decoder circuit with an AND-gate and NOT-gate connected through the programmable switches matrix. The number of AND-gates and NOT-gates depend on the number of input lines to one particular PAL block.

ii) Macrocell block

A macrocell block includes a fixed OR-plane along with an XOR-gate, flip-flop (F/F), multiplexers, and tri-state buffers. Figure 9.6 presents a general view of PAL block with a macrocell. The role of an XOR gate is to obtain the associated complemented form of actual logic function. F/F is used to store the received logic information. A selection between stored and current state is possible with a multiplexer followed by a tri-state buffer at the output stage.

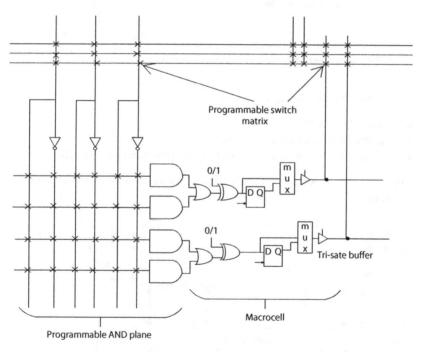

Figure 9.6 PAL-macrocell.

9.3 Field-Programmable Gate Array

A field-programmable gate array (FPGA) is an FPD with a greater number of input/output pins and hardware components along with a very high logic capacity. FPGA provides configurable logic blocks (CLBs) in place of an

AND-OR plane in CPLDs. FPGAs also facilitate a good number of F/Fs and logic resources in comparison to CPLDs [1]. Figure 9.7 shows a general view of an FPGA block diagram. FPGAs are useful in reconfigurable computing where the computer has the software and hardware interfaces with flexible high-speed processing.

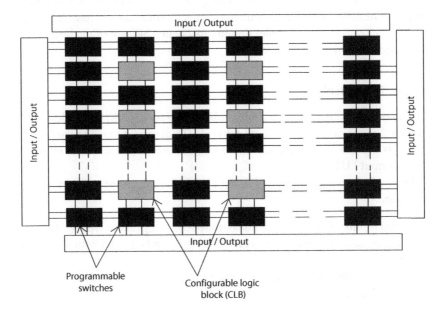

Figure 9.7   FPGA block diagram.

FPGA can handle larger circuits such as:

- CLBs
- I/O blocks
- Interconnection wires and switches
- CLBs that provide logic functionality
- Interconnection switches used to connect CLBs to I/O pins and other blocks of FPGA.

FPGAs contain an array of programmable logic blocks, and a hierarchy of reconfigurable interconnections that allow the blocks to be "wired together," like many logic gates that can be interwired in different configurations. Logic blocks can be configured to perform complex combinational functions, or merely simple logic gates such as AND and XOR. In most FPGAs, logic blocks also include memory elements, which may be simple F/Fs or more complete blocks of memory.

The FPGA is an integrated circuit that contains many (64 to over 10,000) identical logic cells that can be viewed as standard components. Each logic cell can independently take on any one of a limited set of personalities. The individual cells are interconnected by a matrix of wires and programmable switches. A user's design is implemented by specifying the simple logic function for each cell and selectively closing the switches in the interconnect matrix. The array of logic cells and interconnect form a fabric of basic building blocks for logic circuits. Complex designs are created by combining these basic blocks to create the desired circuit.

9.3.1 FPGA Architecture

The major block of FPGA is a configurable logic block (CLB) that is designed with a 2-input or 3-input look up table (LUT) depending on the particular FPGA board. Several commercial FPGA boards are available with four or more than 4-input LUTs [2].

Logic blocks are implemented using LUTs. The important components of LUTs are:

- a fixed number of inputs with one output depending on the type of LUT.
- that they contain storage registers or cells that can be loaded with the suitable values.
- multiplexer blocks to implement any desired function of n-variables depending on the type of LUT.

9.4 Shannon's Expansion and Look-up Table

The Shannon expansion used here is to decide the suitable values to be applied to registers to obtain any desired function from LUTs. Here, an example with two variable functions is illustrated using the Shannon expansion. Suppose a function f has to be implemented in terms of x_1 and x_2 with the following expression:

$$f = x_1 x_2 + x_1 x_2$$

Using Shannon's expansion,

$$f = x_1 \left(x_2 \right) + x_1 \left(x_{2'} \right)$$
$$= x_1 \left(x_2 \right) + x_1 \left(x_{2'} \right)$$

from the above expression, the suitable values for storage register are 0110 which are an output of a 2-input XOR gate. In a similar way, other logic implementation can also be performed using the Shannon expansion method. For example, in f = x'y, the suitable register values for a 2-input LUT will be 0100.

9.4.1 2-Input LUT

A 2-input LUT consists of three 2 × 1 multiplexer blocks with four registers to store suitable values for any desired 2-variable function. Figure 9.8 shows a general block diagram of a 2-input LUT.

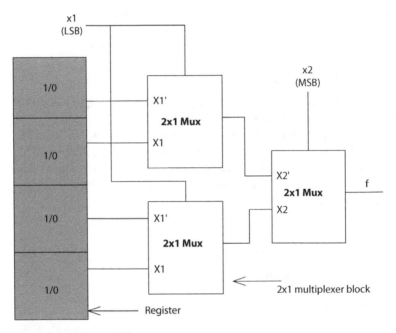

Figure 9.8 2-Input LUT.

Consider an example to implement a function f

$$F = x1 \times x2' + x3 \times 4$$

Here, we have a 4-variable to implement using a 2-input LUT. So, this can be implemented by using three 2-input LUTs. Assume x1.x2'(=f1, 2-variable) and x3x4(=f2, 2-variable),

then f = f1 + f2 can easily be implemented using a 2-input LUT. In a similar manner, any n-variable function implementations are possible with 2-input LUTs. Table 9.1 shows the minimum number of LUTs required along with suitable register values for 2 or >2 variable function implementations using FPGA.

Table 9.1 Examples of function implementation using a 2-input LUT.

Function(f)	No. of 2-input LUT	RegisterValues to LUTs
$x + y$	1	0111
Xy	1	0001
x'y'	1	1000
$(x + y)'$	1	1000
x'y' + xy	1	1001
x1x2 + x3x4	3	0001, 0001, 0111
(x1 + x2)(x3 + x4)	3	0111, 0111, 0001

9.4.2 3-Input LUT

A 3-input LUT consists of seven 2×1 multiplexer blocks with eight registers to store suitable values for any desired 3-variable function.

A general 3-input LUT block is presented in Figure 9.9. For example: f = x1x2x3, the suitable register values will be 00000001 for the 3-input LUT.

The number of storage registers is decided through the number of input variables (equal to 2^n for n-variables) that further decide the multiplexer blocks and their connections.

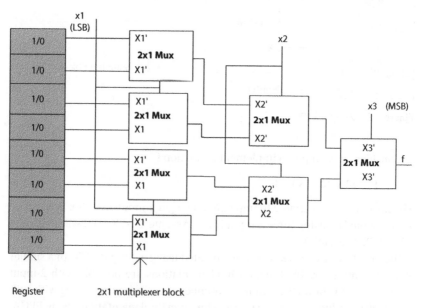

Figure 9.9 3-Input LUT.

9.5 FPGA Families

Several types of product are available called FPGA families or series with different available features, hardware elements, sensors, displays, etc. Xilinx, Inc. Altera Corp. Atmel, and Lattice Semiconductor are but a few companies producing SRAM-based FPGA board. Actel Corp. and Quick Logic Corp are a couple of well-known flash and antifuse FPGAs. Table 9.2 shows a few examples of FPGA families by Xilinx showing their features and technologies.

Table 9.2 Xilinx FPGA family.

Family name	Technology/specifications	Features
XC3000, XC4000, XC5200	0.5 μm–0.25 μm	Old family less hardware and I/O ports
Virtex, Virtex-E, Virtex-EM Virtex-II, Virtex-II PRO, Virtex-4	0.22 μm–90 nm	High-performance family
Basys3 Artix-7 XC7A35T FPGA Nesys4 Artis-7 XC7A100T FPGA	Operating voltage level 1.2 to 3.3 V with 6-input LUT	High-capacity FPGA with onboard peripherals like external temperature sensor, MEMS microphone, accelerometer, I/O devices, along with external memories, collection of USB, ethernet, etc.
Spartan/XL –derived fromXC4000,Spartan-II– derived from Virtex, Spartan-IIE – derived from Virtex-E, Spartan-3	–	Low cost and useful for few specific projects or academic purpose

9.6 Programming with FPGA

Any digital circuit design will be done on an FPGA board using either Verilog HDL or VHDL using Xilinx ISE or Vivado design suit. There is a different version of Xilinx software that is compatible with FPGA-board implementations. The programming FPGA with Xilinx ISE or the Vivado design suit require the following important files:

- HDL source file
- HDL test stimulus
- Xilinx Design Constraint file (UCF or XDC file)

HDL source and test stimulus are common files that describe specific digital-design hardware elements, internal connection and test-input verifications. Master UCF or XDC files are additional important files that give information about the planning for physical onboard pins through software. The Verilog or VHDL code will remain the same for both Xilinx ISE or Vivado design suite for the FPGA board with the only difference being in the design constraint files.

FPGA is an integrated circuit designed to be configured by the designer after manufacturing. It is basically used for rapid prototyping of digital logic block/systems. Figure 9.10 presents a block diagram digital design flow with HDL and an FPGA tool kit:

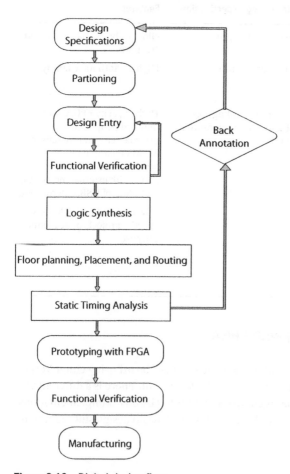

Figure 9.10 Digital design flow.

Design Specifications: Designer is provided with certain design specifications as per standards and requirements such as power, area delay, etc.

Design Partitioning: Designers are allowed to carry out partitioning of the design for their flexibility and simplicity.

Design Entry: This is used to enter the design into an ASIC design system, either using hardware description language (HDL) or schematic entry.

Functional Verification: In this step, the circuit design is verified, i.e., the system functioning according to the desired specifications. If not, back annotation is done and design entry is changed here. Static Timing Analysis Prototyping on the FPGA Batch Process (Manufacturing) Functional Verification Specifications System Partitioning (For Large Systems) Design Entry Functional Verification Logic Synthesis Floor Planning Placement and Routing Back Annotation by the designer.

Prototyping on FPGA: When post-layout simulation is done correctly, then a prototype is designed by burning the software code on FPGA.

Manufacturing: After designing a prototype, the chip will be manufactured according to the designed prototype.

9.6.1 Introduction to Xilinx Vivado Design Suite for FPGA-Based Implementations

Different versions and packages of the Vivado design suite are freely available with an FPGA tool kit. In a Vivado design suite, there are project-mode platforms for creating new projects, design-source files, design analysis & synthesis, bit-stream generation, TCL Script creation, and implementation on FPGA. FPGA is most commonly used in the area of aerospace, defenseportable medical electronics, ASIC Prototyping, Digital Signal Processing (DSP), high-resolution video, image processing and consumer electronics such as displays, switches, digital cameras, multi-function printers, etc.

9.7 ASIC and Its Applications

The reconfigurable capability of FPGA makes it more popular in VLSI design and application. But still the need for ASIC will always be there where an IC has been designed to perform a specific task or application. ASIC is basically designed with optimum use of the hardware element so it is less costly when manufactured in bulk. Also, the power consumption, delay and IC area are less in comparison to a similar design on the FPGA board. Sometimes, FPGA boards are utilized to design and analyze ASIC before fabrication processes to save time and cost involved in fabrication. The design failures of ASIC following fabrication can be reduced beforehand by taking such verification steps with FPGA.

Review Questions

Q1 Compare SPLD with CPLD in terms of hardware components and applications.

Q2 What is ASIC? How do ASICs differ from FPGAs?

Q3 How is FPGA is advantageous over SPLD and CPLDs?

Q4 What is LUT? Give an example of a 2-input LUT.

Q5 What do you mean by CLB? Give one example of CLB.

Q6 Use a CPLD to implement the function

$$f = x1.x3.x6' + x1.x4.x5.x6' + x2.x3.x7 + x2.x4.x5.x7$$

Q7 Use the Shannon expansion to implement the following function:

a) $f1 = x_1x_2$ b) $f2 = x1'x2$ c) $f3 = f1 + f2$

Multiple Choice Questions

Q1 The correct statement for LUT:
  A LUT can be elaborated as a look-up table
  B It can be designed with the help of 2-input or 3-input multiplexer blocks
  C Any logic function can be implemented by applying suitable input values to a look-up table
  D All of the above

Q2 ASIC is different from FPGA because:
  A ASIC is an analog integrated circuit while FPGA is a digital integrated circuit

B ASIC is an application specific integrated circuit while FPGA can be programmed for different functionality

C FPGA is designed for specific application while ASIC can be programmed for different functionality

D None of the above

Q3 Select a suitable input value for a 2-input LUT for the given function $f = a'b$

A 0100

B 1100

C 1000

D 0001

Q4 The correct expression after Shannon expansion for the given function $f = x_1'x_2' + x_1x_2$ is?

A $x_1'(x_2') + x_1(x_2)$

B $x_1'(x_2'(1) + x_2(0)) + x_1(x_2'(0) + x_2(1))$

C $x_1'(x_2'(0) + x_2(0)) + x_1(x_2'(0) + x_2(1))$

D $x_1'(x_2'(1) + x_2(1)) + x_1(x_2'(0) + x_2(1))$

Q5 How many input registers will be there in one block of 3-input LUT of FPGA?

A 2

B 4

C 6

D 8

Q6 The CPLD architecture includes:

A PAL, I/O block, Interconnecting wires

B LUT, I/O block, Interconnecting wires

C Configurable Logic Block, Interconnecting switches, I/O block, Interconnecting wires

D PLA, I/O block, Interconnecting wires

Q7 The FPGA architecture includes:

A PAL, I/O block, Interconnecting wires

B PLA, I/O block, Interconnecting wires

C Configurable Logic Block, Interconnecting switches, I/O block, Interconnecting wires

D None of the above

Q8 Select a suitable input value for a 2-input LUT for the given function
$f = f1.f2$ where $f1 = (a + b)$, $f2 = (a + b)$

A 0001

B 0111

C 1000

D 1001

Q9 Select the suitable example for FPGA series:

A XC3000

B Virtex

C Spartan

D All of the above

Q10 PAL blocks are a combination of:

A programmable OR gate and fixed AND gate

B programmable AND gate and fixed OR gate

C both programmable AND, OR gates

D both fixed AND, OR gates

Q11 How many 2-input LUTs are required to implement the given function
$f = (x_1x_{6'} + x_2x_7)(x_3 + x_4x_5)$?

A 4

B 5

C 6

D 7

Q12 The major difference between SPLD and CPLD is?

A SPLD has PAL block and CPLD has PLA block

B SPLD has PLA block and CPLD has PAL block

C The total number of n(input + output) < 100 in SPLD while
n > 100 in CPLD

D The total number of n(input + output) < 32 in SPLD while n > 32
in CPLD

Q13 How many 2-input LUTs are required to obtain the given function
$f = x_1x_2 + x_{2'}x_3$?

A 2

B 3

C 4

D 5

Q14 FPGAs are most popular programmable logic devices because:
  A They are perfect for rapid prototyping of digital circuits
  B They support reconfigurable computing
  C They are easy upgrades such as in the case of software
  D All of the above

References

[1] Unsalan, C., and Tar, B. (2017). *Digital System Design With FPGA: Implementation using Verilog and VHDL*. New York, NY: McGraw Hill Education.

[2] Wolf, W. (2004). *FPGA-Based System Design*. Upper Saddle River, NJ: Pearson.

10

Project Based on Verilog HDLs

This chapter includes VLSI projects based on digital circuit design using Verilog programming and functional verification with a truth table on Xilinx tool [1–2]. The project includes all four levels of abstraction of Verilog from the switch-level to behavioral-level model. Each project gives a basic description of design including truth table and design verification results.

Xilinx ISE is one of the useful simulators that uses Verilog/VHDL languages to design and implement any digital logic virtually. Figures 10.1–10.4 show how the Xilinx ISE simulator interfaces to create a project, select the Verilog module, text editor, and waveform window, respectively.

Step 1: file → new project → project name

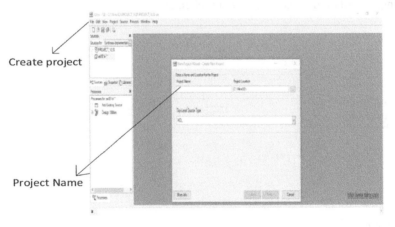

Figure 10.1   New project creation on Xilinx ISE simulator.

Digital VLSI Design and Simulation with Verilog, First Edition. Suman Lata Tripathi, Sobhit Saxena, Sanjeet Kumar Sinha, and Govind Singh Patel.
© 2022 John Wiley & Sons Ltd. Published 2022 by John Wiley & Sons Ltd.

Step 2: Right click on project created → new source → select Verilog module → module name. Choose option next and finish.

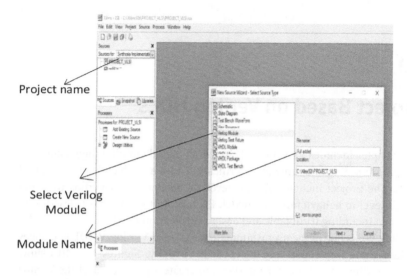

Project name

Select Verilog Module

Module Name

Figure 10.2 New source module creation on Xilinx.

Step 3: Write a Verilog program using the platform of text editor. Select synthesis implementation for getting a circuit schematic.

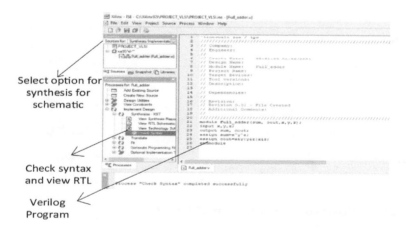

Select option for synthesis for schematic

Check syntax and view RTL

Verilog Program

Figure 10.3 Xilinx platform for Verilog HDL.

Step 4: Write the test stimulus and select the option of behavioral simulation to obtain waveform and to verify the truth table.

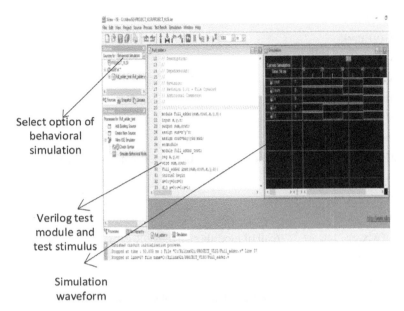

Select option of
behavioral
simulation

Verilog test
module and
test stimulus

Simulation
waveform

Figure 10.4   Behavioral simulation on Xilinx platform.

10.1   Project Based on Combinational Circuit Design Using Verilog HDL

Any digital circuit can be designed on the Xilinx ISE platform using different levels of abstraction of Verilog HDL. Combinational circuits such as adder, subtractor, multiplier, comparators, etc. are usually part of many digital processors. In this section, a few additional examples of combinational circuit are presented using switch-level, gate-level, dataflow, and behavioral models.

10.1.1   Full Adder Using Switches at Structural-Level Model

In this example, a full-adder module has been designed using switch-level modules of logic gates AND, OR and XOR etc. The design consists of four modules including the top full-adder module with submodules of switches such as *snot*, *sxor*, *sor*, and *sand*. The switch-level implementation of logic gates at transistor level are discussed in Chapter 7.

//Main module

module *Switch_fulladd*(S, C, u, v, w);   //Top module and port declaration

input u,v,w;

output S, C;

```
wire y1, c1, c2;
sxor u1(y1,u,v);
sxor u2(S,y1,w);
sand u3(c1,u,v);
sand u4(c2,y1,w);
sor u5(C, c2,c1);
endmodule
module sor(f,x,y);          //module elaboration of OR-gate using switch (sor)
input x,y;
output f;
wire w;
snor x1(w,x,y);             //call of module switch NOR-gate
snot x2(f,w);
endmodule
module snor(Op,x,y);        //module elaboration of switch nor with port
declaration
output Op;
input x,y;
//internal wires
wire w;
supply1 Vdd;                //Power supply is connected to Vdd
supply0 Gnd;                //Ground is connected to Gnd
pmos (w, Vdd, y);
pmos (Op, w, x);
nmos (Op, Gnd, x);
nmos (Op, Gnd, y);
endmodule
module sxor(Op,x,y);
input x,y;
output Op;
wire w1,w2,w3,w4,w5;
supply1 Vdd;
supply0 Gnd;
nmos (w3,Gnd,w2);
nmos (Op,w3,w1);
nmos (w4,Gnd,y);
```

```
nmos (Op,w4,x);
pmos (Op,w5,x);
pmos (Op,w5,y);
pmos (w5,Vdd,w1);
pmos (w5,Vdd,w2);
snot n1(w1,x);          //call of module switch NOT-gate
snot n2(w2,y);
endmodule
module sand(Op,x,y);    //module elaboration of AND-gate using switch (sand)
output Op;
input x, y;
wire w;
snand x1(w,x,y);
snot x2(Op,w);
endmodule
module snand(Op,x,y);   //module elaboration of NAND-gate using switch
                        (snand)
output Op;
input x,y;
wire c;
supply1 Vdd;
supply0 Gnd;
pmos (Op,Vdd, y);
pmos (Op,Vdd, x);
nmos (w, Gnd, x);
nmos (Op, w, y);
endmodule
module snot(Op,x);      ////module elaboration of NOT-gate using switch (snot)
input x;
output Op;
supply1 Vdd;
supply0 Gnd;
nmos (Op,Gnd,x);
pmos (Op,Vdd,x);
endmodule
```

10.1.2 Ripple-Carry Full Adder (RCFA)

A 4-bit RCFA has been designed with four blocks of 1-bit full adder using the structural-level model (Figure 10.5). A 1-bit full-adder module can be elaborated as a submodule using the gate or dataflow model which is discussed in Chapter 4. Variables M & Cout are considered 4-bit and 1-bit size output while U & V are considered 4-bit input variables. Cin is taken for initial carry input for the block of 4-bit RCFA.

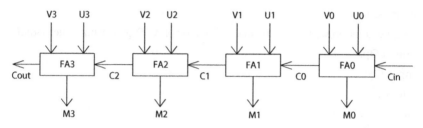

Figure 10.5 4-bit ripple-carry full adder.

//Design module

module RCFA_4_bit (M,Cout,U,V,Cin);

input [3:0] U,V;

input Cin;

output [3:0]M;

output Cout;

wire C0,C1,C2;

fulladd f1(M[0],C0,U[0],V[0],Cin);

fulladd f2(M[1],C1,U[1],V[1],C0);

fulladd f3(M[2],C2, U[2],V[2],C1);

fulladd f4(M[3],Cout, U[3],V[3],C2);

endmodule

10.1.3 4-bit Carry Look-ahead Adder (CLA)

CLAs come under the category of fast adders. Here, the full-adder (FA) block need not wait for the carry generated from the previous block of FA. However, it takes more hardware logic block compared to an RCFA. Figure 10.6 presents a 4-bit CLA block. The design of a CLA block is done using the dataflow model on a Verilog HDL platform.

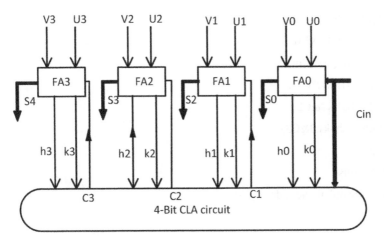

Figure 10.6 4-bit CLA adder.

//Design module

```
module carrylookahead_adder(S,Cout,U,V,Cin);
input [3:0]U,V;
input Cin;
output [3:0]S;
output Cout;
wire h0,k0,h1,k1,h2,k2,h3,k3;
wire C3,C2,C1;
assign k0 = U[0]^V[0];
assign k1 = U[1]^V[1];
assign k2 = U[2]^V[2];
assign k3 = U[3]^V[3];
assign h0 = U[0]&V[0];
assign h1 = U[1]&V[1];
assign h2 = U[2]&V[2];
assign h3 = U[3]&V[3];
assign S[0] = U[0]^V[0]^Cin;
assign S[1] = U[1]^V[1]^C1;
```

```
assign S[2] = U[2]^V[2]^C2;
assign S[3] = U[3]^V[3]^C3;
assign C1 = h0|(k0&Cin);
assign C2 = h1|(k1&C1);
assign C3 = h2|(k2&C2);
assign Cout = h3|(k3&C3);
endmodule
//Test Stimulus
module carrylookahead_adder_test;
reg [3:0]U,V;
reg Cin;
wire [3:0]S;
wire Cout;
carrylookahead_adder u(S,Cout,U,V,Cin);
initial begin
$monitor ($time,"S = %d,Cout = %d,U = %b,V = %b,Cin =
%b",S,Cout,U,V,Cin);
U = 4'd1;V = 4'd2;Cin = 0;
#15 U = 4'd2;V = 4'd5;Cin = 0;
#15 U = 4'd4;V = 4'd4;Cin = 1;
#15 U = 4'd3;V = 4'd5;Cin = 0;
#20 $stop;
end endmodule
```

10.1.4 Design of a 4-bit Carry Save Adder (CSA)

A 4-bit carry save adder is designed at the structural level using gate-level modules of a 1-bit full adder as shown in Figure 10.7. CSAs are mainly used for the addition of three or more than three binary numbers such as multipliers where the addition of more than two binary numbers is frequently performed in multiplication operations. Here, a full-adder module has been called to implement CSA that can be declared using either the gate or dataflow level in a separate module.

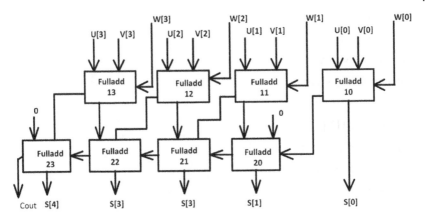

Figure 10.7 4-bit CSA block diagram.

```
//Main module
module CSA(u,v,w,s,cout);
input [3:0] u,v,w;
output [4:0] s;
output cout;
wire [3:0] c1,s1,c2;
fulladd fa_10(u[0],v[0],w[0],s1[0],c1[0]);      //calling of full blocks
fulladd fa_11(u[1],v[1],w[1],s1[1],c1[1]);
fulladd fa_12(u[2],v[2],w[2],s1[2],c1[2]);
fulladd fa_13(u[3],v[3],w[3],s1[3],c1[3]);
fulladd fa_20(s1[1],c1[0],1'b0,s[1],c2[1]);
fulladd fa_21(s1[2],c1[1],c2[1],s[2],c2[2]);
fulladd fa_22(s1[3],c1[2],c2[2],s[3],c2[3]);
fulladd fa_23(1'b0,c1[3],c2[3],s[4],cout);
assign s[0] = s1[0];
endmodule
```

10.1.5 2-bit Array Multiplier

```
module arraymultiplier_2bit(C,X,Y);
input [1:0]X,Y;
```

```
output [3:0]C;
wire w1,w2,w3,w4;
and x1(C[0],X[0],Y[0]);
and x2(w1,X[0],Y[1]);
and x3(w2,X[1],Y[0]);
and x4(w4,X[1],Y[1]);
ha h1(C[1],w3,w1,w2);
ha h2(C[2],C[3],w3,w4);
endmodule
module ha(s,c,x,y);
input x,y;
output s,c;
xor r1(s,x,y);
and r2(c,x,y);
endmodule
module arraymultiplier_2bit_test;
reg [1:0]x,y;
wire [3:0]C;
arraymultiplier_2bit u(C,x,y);
initial begin
x = 2'd0;y = 2'd0;
#5 x = 2'd0;y = 2'd1;
#5 x = 2'd1;y = 2'd0;
#5 x = 2'd2;y = 2'd3;
#5 $stop;
end endmodule
```

10.1.6   2 × 2 Bit Division Circuit Design

A 2 × 2 bit division circuit has been designed with the help of a truth table and K-map simplification. Here, the simple rule of division is followed as: $0/0 = x;0/1 = 0;1/1;1/0 = x$ etc. Figure 10.8 illustrates the truth table and K-map for simplified output expressions.

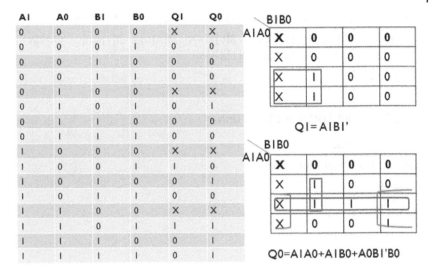

A I	A0	B I	B0	Q I	Q0
0	0	0	0	X	X
0	0	0	I	0	0
0	0	I	0	0	0
0	0	I	I	0	0
0	I	0	0	X	X
0	I	0	I	0	I
0	I	I	0	0	0
0	I	I	I	0	0
I	0	0	0	X	X
I	0	0	I	I	0
I	0	I	0	0	I
I	0	I	I	0	0
I	I	0	0	X	X
I	I	0	I	I	I
I	I	I	0	0	I
I	I	I	I	0	I

Q I = A I B I '

Q0 = A I A0 + A I B0 + A0B I 'B0

Figure 10.8 Truth table and K-map.

//**Main module**

module division2x2 (Q,A,B);

input [1:0]A,B;

output [1:0]Q;

assign Q1 = A1& ~ B1;

assign Q2 = (A1&A0)|(A1&B0)|(A0B1'B0);

endmodule

10.1.7 2-bit Comparator

In this example, a 2-bit comparator circuit is designed with the help of truth table and dataflow models. Here, f1, f2, and f3 are used to show the relation between A < B, A > B, and A = B, respectively. The expression is obtained through K-map.

//Main module

module comparator_2bit(f1,f2,f3,A,B);

input [1:0]A,B;

output f1,f2,f3;

assign f1=(~A[1]&B[1])|(~A[1]&~A[0]&B[0])|(~A[0]&B[1]&B[0]);

assign f2=(A[0]&~B[1]&~B[0])|(A[1]&~B[1])|(A[1]&A[0]&~B[0]);

assign f3=(~A[1]&~A[0]&~B[1]&~B[0])|(~A[1]&A[0]&~B[1]&B[0])|(A[1]&~A[0]&B[1]&~B[0])|(A[1]&A[0]&B[1]&B[0]);

endmodule

10.1.8   16-bit Arithmetic Logic Unit

```
//Design module
module ALU_16bit(out,a,b,s);
input [15:0]a,b;
input [3:0]s;
output reg [31:0]out;
always@(s)
begin
case(s)
4'd0:out=a+b;
4'd1:out=a*b;
4'd2:out=a&b;
4'd3:out=a^b;
4'd4:out=a<<2;
4'd5:out=b>>2;
4'd6:out={a,b};          //concatenation operator
4'd7:out={2{a}};         //Replication operator
4'd8:out=a-b;
4'd9:out=~a;
4'd10:out=|b;
4'd11:out=a>>1;          //right shift 1 bit
4'd12:out=b<<1;          //left shift 1 bit
4'd13:out=~a^b;
4'd14:out=&a;            //reduction operator
4'b15:out=~(a|b);
endcase
end endmodule
```

10.1.9 Design and Implementation of 4 × 16 Decoder Using 2 × 4 Decoder

Decoders are part of various circuits that have multiple input and output lines. In this example, a 4 × 16 decoder circuit has been designed with the help of submodules of 2 × 4 decoder using structural-level and dataflow models. The outputs of the decoder in the first stage are used to enable the decoders in the second stage. Figure 10.9 illustrates the logic block for implementation of 4 × 16 decoder design using 2 × 4 decoder blocks.

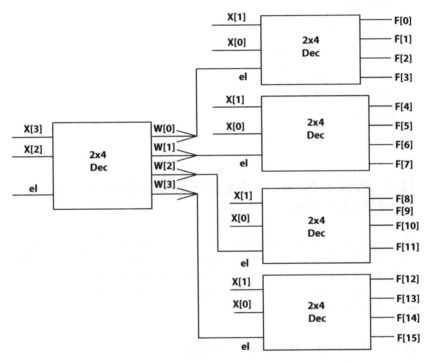

Figure 10.9 1.8: 4 × 16 decoder using a 2 × 4 decoder.

//Design modules

module decoder_4x16(f,el,x);

output [15:0] f,

input [3:0] x,

input el;

wire [3:0] w;

dec_2x4 r0(w, x[3:2], el);

dec_2x4 r1(f[3:0], x[1:0], w[0]);

dec_2x4 r2(f[7:4], x[1:0], w(1));

dec_2x4 r3(f[11:8], x[1:0], w(2));

dec_2x4 r4(f[15:12], x[1:0], w[3]);

endmodule

module dec_2x4 (f0,f1,f2,f3,el,a,b);

input a, b, en;

output f0,f1,f2,f3;

assign f0 = (~a) & (~b) & en;

assign f1 = (~a) & b & en;

assign f2 = a & (~ b) & en;

assign f3 = a & b & en;

end module

10.2   Project Based on Sequential Circuit Design Using Verilog HDL

Similar to a combinational circuit, sequential circuits are also implemented using different levels of abstraction Verilog using logic blocks, F/Fs, and the clock signal. The behavioral-level model is the most preferred for sequential circuit design. In this section, a few examples of sequential circuit implementations have been described.

10.2.1   Design of 4-bit Up/down Counter

Counter is an example of a sequential circuit that counts pulses. In this example, a counter has been designed to perform both up-counting as well as down-counting with the help of the behavioral-level model. Here, the counter block is triggered with the positive edge of the *clock* signal. A *clear* signal is used to clear the values stored in different F/Fs of up-down counters. The input *up_down* is used here to select between the up or down counting process.

//Design module

module asyn_up_down(out, clock, clear, up_down);

input clock, clear, up_down;

output reg[3:0] out;

always@(posedge clock, posedge clear)

begin

if(clear = = 0)

begin

if(up_down = = 1)

out = out + 1;

else out = out-1;

end

else

out = 4′d0;

end

endmodule

10.2.2 LFSR Based 8-bit Test Pattern Generator

Linear feedback shift registers (LFSRs) are used to generate random test input sequences. They are frequently used in built-in-self-test circuits to generate the desired test input patterns. The required number of F/Fs depends on the number of bits in sequence. Here, x1,x2,x3 show different blocks of F/F. Figure 10.10 presents an 8-bit LFSR logic block.

//Design module:2 LFSR

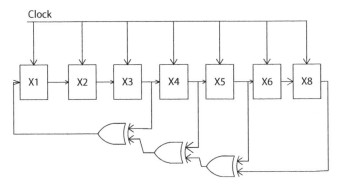

Figure 10.10 8-bit LFSR.

```verilog
//Design module
module lfsr8_xor(out,clk,rst);
output reg [7:0] out;
input clk, rst;
wire xx0,xx1,xx2;
assign xx2 = (out[6] ^ out[7]);
assign xx1 = (out[5] ^ xx0);
assign xx0 = (out[4] ^ xx1);
always @(posedge clk, posedge rst)
begin
if (rst)
out = 8'd00000001;
else
out = {out[6:0],xx2};
end
endmodule
//Test Stimulus
module lfsr_tb();
reg clk;
reg rst;
wire [7:0] out;
lfsr8_xor u(out,clk,rst);
initial begin
clk = 0;rst = 1;
#15 rst = 0;
#200;
end
always #5clk = ~ clk;
end
endmodule
```

10.3 Counter Design

Counter design is easily done with the help of conditional or loop statements at behavioral-level model. *if/else* or *for* loop-based examples are provided here.

10.3.1 Random Counter that Counts Sequence like 2,4,6,8,2,8...and so On

This is an example of random counter with the use of if else statement in Verilog HDL.

```
//Design module
module counter_random(count, clk, rst);
input clk,rst;
output reg [3:0]count;
always@(posedge clk)
begin
if(rst)
count ≤ 4'd0;
else if(count < 10)
count ≤ count + 2;
else
count ≤ 4'd2;
end
endmodule
module counter_random_test;
reg clk,rst;
wire [3:0]count;
counter_random u(count,clk,rst);
initial begin
clk = 0;
forever #5clk = ~ clk;
end
```

```
initial begin
rst = 1;
#10 rst = 0;
#200 $stop;
end
endmodule
```

//Mod16 up-counter that count from 0-15

```
module count_mod16(co,clk,rst);
input clk,rst;
output reg [3:0]co;
integer i;
always@(posedge clk)
begin
if(rst)
co ≤ 0;
else
for(i = 0;i < 15;i = i + 1)          //use of for loop
co ≤ co + 1;
end
endmodule
```

//Mod10 down counter that count from 0-9

```
module counter_down(coclk,rst);
input clk,rst;
output reg [3:0]co;
always@(posedge clk)
begin
if(rst)          //use of if else
co ≤ 0;
else if(co = = 0)
co ≤ 4'd9;
else
```

co ≤ co-1;

end

endmodule

10.3.2   Use of Task at the Behavioral-Level Model

In this example, a logic has been developed to count the number of 1 from input binary string using tasks as subroutine in the behavioral-level model.

```verilog
//Main module
module task_count_1(cout,x);
input [7:0]x;
output reg [3:0]cout;
integer i;
always@(x)
count_1(cout,x);
task count_1;
output [3:0]cout;
input [7:0]x;
begin
cout = 8'd0;
for(i = 0;i < 8;i = i + 1)
if(x[i])            //x[i] is true
cout = cout + 1;
else
cout = cout;
end
endtask
endmodule
//Test bench
module task_count_1test;
reg [7:0]x;
wire [3:0]cout;
```

```
task_count_1 u(cout,x);
initial begin
x = 8'd0;
#5 x = 8'd7;
#5 x = 8'd9;
#5 x = 8'd8;
#5 x = 8'd15;
#5 $stop;
end
endmodule
```

10.3.3 Traffic Signal Light Controller

```
//Design module
module traffic_signal_lights;
reg clk, cred, cyellow, cgreen;
parameter on = 1, off = 0, cred_tics = 30,
cyellow_tics = 3, cgreen_tics = 20;
initial cred = off;
initial cyellow = off;
initial cgreen = off;
always begin
cred = on;
light(cred, cred_tics);
cgreen = on;
light(green, green_tics);
cyellow = on;
light(cyellow, cyellow_tics);
end
//task to wait for positive edge clocks of tic
task light;
output col;
```

```
input [31:0] tic;
begin
repeat (tic) @ (posedge clk);
col = off;
end
endtask
always begin
#15 clk = 0;
#1 clk = 1;
end
endmodule
```

10.3.4 Hamming Code(h,k) Encoder/Decoder

Hamming code is one of the techniques to check and correct error bit from an input bit stream. It can be designed in blocks named as generator, checker, and corrector circuit.

```
//Encoder:
module ham_en(clk,i,co);
input clk;
input [3:0] i;
output reg[6:0] co;
always@(posedge clk)
begin
co[6]=i[3];
co[5]=i[2];
co[4]=i[1];
co[3]=i[1]^i[2]^i[3];
co[2]=i[0];
co[1]=i[0]^i[2]^i[3];
co[0]=i[0]^i[1]^i[3];
end
```

endmodule

//Decoder

module ham_dec(co,clk,w,cout,d);

input clk;

input[6:0] co;

output reg[2:0]w;

output reg[6:0] cout;

output reg[3:0]d;

always@(posedge clk)

begin

w[2]=co[0]^co[4]^co[5]^co[6];

w[1]=co[1]^co[2]^co[5]^co[6];

w[0]=co[0]^co[2]^co[4]^co[6];

cout=co;

if(w)

cout[w-1]=~c[w-1];

end

always@(cout)

begin

d[0]=cout[2];

d[1]=cout[4];

d[2]=cout[5];

d[3]=cout[6];

end

endmodule

Frequency Divider Circuit

A frequency divider is useful to increase clock pulse duration by frequency divison f/n, where n = 1,2,3...so on. As per the expression of clock time period, $T = \dfrac{1}{f} = \dfrac{1}{1MHz} = 1\mu sec$ for 1 MHz frequency. Such a high-frequency clock pulse cannot be observed by the human eye because of the very low time period. For FPGA-based high-frequency implementation, a frequency divider circuit is required to observe the response in terms of input and output variations with increased delay as per required. A Verilog program for frequency divider has been illustrated below for frequency division circuit $f/2^{25}$ for the frequency above 100 MHz.

```
//Design Module
module clockdivide(clk, nclk);        //clk and nclk are the clock input and
                                      output of divider circuit
input clk;
output reg nclk;
reg [31:0]count = 32'd0;
always@(posedge clk)
begin
count = count + 1;
nclk = count[25];
end
endmodule
//Test Stimulus
module clockdivide_test;
reg clk;
wire nclk;
clockdivide uu(clk, nclk);
initial begin
clk = 0;
forever #5clk = ~ clk;
end
initial
# 1000000000 $stop;
endmodule
```

Review Questions

Q1 Design and write a Verilog program for 32 × 1 multiplexer using 4 × 1 multiplexer.

Q2 Write a Verilog program of decoder circuit to drive seven-segment display.

Q3 Write a Verilog program for frequency divider to increase with increased clock pulse delay by 4 times.

Q4 Write a Verilog program of 2-to-4 line decoder using switches.

Q5 Write a Verilog program of 4-bit adder/subtractor circuit using the data-flow model.

Multiple Choice Questions

Q1 Which statement is true for test stimulus in Verilog HDL?
  A The stimulus block instantiates the design block and directly drives the signals in the design block
  B Stimulus is to instantiate both the stimulus and design blocks in a top-level dummy module
  C The stimulus block interacts with the design block only through the interface
  D All of the above

Q2 Choose the correct syntax of 4 × 1 multiplexer using conditional operator
  A assign Y = S1?(S0?d3:d2):(S0?D1:D0);
  B assign Y = S1?(S0?d3:d2):(S1?D1:D0);
  C assign Y = S0?(S1?d3:d2):(S0?D1:D0);
  D None of the above

Q3 Which is the correct syntax for the implementation of Y(a,b) = $\sum m(0,2)$?
  A assign Y = (~a| ~ b)&(a| ~ b);
  B assign Y = (~a& ~ b)|(a& ~ b);
  C assign Y = (~a&b)|(a& ~ b);
  D assign Y = (a& ~ b)|(a& ~ b);

Q4 For frequency divider f/4, the clock period will be:
    A T/4
    B 4T
    C T
    D None

Q5 For 100 MHz frequency, time period will be:
    A 0.1μsec
    B 0.01μsec
    C 0.001msec
    D none

References

[1] XILINX. (2021). Website. https://www.xilinx.com (accessed 22 June 2021).
[2] Man, M. and Ciletti, M.D. (2013). *Digital Design: With an Introduction to the Verilog HDL*, 5e. ed. Upper Saddle River, NJ: Pearson.

11

SystemVerilog

11.1 Introduction

After studying Verilog in detail, let's go one step further and learn about SystemVerilog. Due to some additional features compared to Verilog, this becomes attractive for fast-evolving technologies and industries are now more interested. SystemVerilog obtained standardization by the IEEE 1800 in 2018 as a hardware description and hardware verification language [1] which is an extension of Verilog used to model, design, simulate, test, and implement electronic systems. For better understanding, the features of SystemVerilog are discussed in comparison with Verilog.

11.2 Distinct Features of SystemVerilog

There are many extended features that make the SystemVerilog a better hardware designing and verification language than Verilog. A few of these are:

(i)   Non-blocking and blocking operators ($\leq$ and $=$ respectively) can be used for arrays.
(ii) Input, output, and inout ports support more formats of data types such as real, struct, and enum. Multi-dimensionality is also supported.
(iii) Equipped with the automatic declaration of variable inside the loop statement. Addition of do/while loop in the while-loop construct.

Digital VLSI Design and Simulation with Verilog, First Edition. Suman Lata Tripathi, Sobhit Saxena, Sanjeet Kumar Sinha, and Govind Singh Patel.

(iv) Many new operators similar to C-language are supported in SystemVerilog. A few of these are:

    a) Increment/decrement operators viz. i++, ++i, i–, –i

    b) Compound-assignment operators viz. i + = x, i- = x, i* = x, i/ = x, i% = x, i≪ = x, i≫ = x, i& = x, i^ = x, i| = x.

(v) New features in the fork-join block have been added viz. join_none & join_any.

(vi) It is now possible in SystemVerilog that Functions return no value and are declared as void.

(vii) Parameters can be declared of any type, including user-defined typedefs.

(viii)There is a facility in SystemVerilog for interfacing with other languages such as C&C++ known as Direct Programming Interface (DPI).

There are so many things that can be done using SystemVerilog. In the upcoming sections, the detailed description of SystemVerilog in terms of data types, interfaces, clocking, classes, etc. is available which will make you comfortable when programming with SystemVerilog.

11.2.1 Data Types

SystemVerilog introduced new data types in almost all categories. New data types are inspired from the C-language which makes programmers more comfortable when switching from C to SystemVerilog.

> **Two-state variable type:** Compared to variable in Verilog which has four possible states 0, 1, X, and Z, the variable in SystemVerilog can have two states; 0 and 1 only. For example, bit, byte, shortint, int and longint.
>
> **"logic" in place of "reg" or "wire":** SystemVerilog gives the freedom from deciding which one needs to declare as **reg** and which as **wire**. The SystemVerilog programmer only has to declare **logic** and it is the job of a synthesis tool to convert it into reg or wire as per the requirement of design.

Example 1: Difference in declaring type of input/output and endmodule syntax.

The SystemVerilog code for a 4-bit adder is shown below followed by the same program coded in Verilog for comparison. The main differences between these two codes are highlighted in bold.

module adder4bit (a,b,c,sum,carry);

input **logic** [3:0]a,b;

input **logic** c;

output **logic** [3:0]sum;

output **logic** carry;

logic [4:0] result;

assign result = a + b + c;

assign sum = result [3:0];

assign carry = result[4];

endmodule:**adder4bit**

module adder4bit (a,b,c,sum,carry);

input reg [3:0]a,b;

input reg c;

output wire [3:0]sum;

output wire carry;

wire [4:0] result;

assign result = a + b + c;

assign sum = result [3:0];

assign carry = result[4];

endmodule

There is one difference in the syntax of module and endmodule i.e., the module name is also written after endmodule separated by a colon ":". The other difference is declaration of "logic" instead of "wire" as well as "reg." If we declare signals as "logic" in SystemVerilog, the synthesizing tool will sort out if it is a "wire" or "reg."

11.2.2 Arrays

In Verilog, though multi-dimensioned arrays of both nets and variables are allowed, some of the restrictions are still there on memory array usage. Unpacked and packed dimension arrays in SystemVerilog allow more operations. Packed dimensions are declared before the name and unpacked after the name.

Example 1: reg [3:0][4:0] register [0:6];
where [3:0] and [4:0] are packed dimensions while [0:6] is an unpacked dimension. Any number of packed/unpacked dimensions can be declared without any limit.

Packed dimensions are useful in designing the memory in personal style. Systematic approaches make the memory closely packed. Slicing i.e., part-select can be done more effectively. Packed dimension objects can be copied onto any other packed object. There is a restriction on the data type that only "bit" types are allowed.

In contrast with packed dimensions, the arrangement of unpacked dimensions in memory can be done in any way depending upon the choice of simulator. Copying can be done reliably from one array to another array of the same type. If the arrays are of different types, there is a facility of cast available by which any unpacked array can be converted to a packed array by following a set of rules. There is a restriction on the data type in unpacked arrays, for example, real is also allowed.

If the shape and type of unpacked arrays and their slices are the same, then different operations can be performed in SystemVerilog. Same shape and same type here means the number and lengths of unpacked-dimensions should be exactly the same. If the number of bits in the array and slice elements are different, packed-dimensions are allowed. These operations include writing and reading the complete array, array slices, and array elements. Equality relations can also be applied on arrays, slices, and elements.

Example 2: Two-dimensional declaration of ports
The two-dimensional array data type is supported by Verilog but that data type is not allowed for use in declaration of the ports. The possibility to do the same in Verilog is that, first, the port is declared as a one-dimensional array and afterwards a two-dimensional signal will be generated internally. One example of this is shown below.

module two_dimen_verilog

(. .

//input reg [15:0] aone [31:0] not allowed in Verilog

input reg [32*16-1:0] a_one,

. .);

//two dimensional reg can be declared internally

reg [15:0] a_two [31:0];

//generation of two dimension signal inside the module

genvar i;

generate

for (i = 0; i < 32; i = i + 1) begin:

assign a_two[i] = a_one[(i + 1)*16-1: i*16];

end

endgenerate

. . .

Port declaration of a two-dimensional array data type is fully supported in SystemVerilog. The above Verilog code can be rewritten in SystemVerilog as:

module two_dimen_sys_verilog

(. . .

input logic [15:0]a_two [31:0],

. . .);

The concept of dynamic arrays is also available in SystemVerilog. Dynamic arrays allowed the change in a number of elements during simulation. Another type of array with a non-contiguous range, known as associative array, is also available in SystemVerilog. To handle a variety of arrays, some array querying functions and methods are provided in SystemVerilog. For example, the number of dimensions of any array can be identified using "$dimensions" keyword.

11.2.3 Typedef

Sometimes, with the available data types, the complexity of the developed system increases greatly. In SystemVerilog, there is a facility to create names for the corresponding complex data type which is going to be used frequently inside the program. To make this possible, keyword "typedef" is available in SystemVerilog. "typedefs" can be very useful while creating complex array-definitions.

Example 3: typedef usage to define complex data type

typedef reg [3:0] quater;

quater b;

//is same as

reg [7:0] b;

//and

typedef quater [7:0] Octaquater;

Octaquater newbyte [1:12];

//is same as

reg [7:0][3:0] newbyte [1:12];

11.2.4 Enum

The concept of enumeration is also introduced in SystemVerilog. The keyword used for this application is "enum"

e.g., enum {triangle, square, rectangle} s;

Enumerations allow definition of a different type of data-type in which values are defined by names. These data-types are very useful and appropriate for representation of symbolic-data and non-numeric values viz. state-values, opcodes, etc. "typedef" is used with "enum" very commonly,

   e.g.,

typedef enum {triangle, square, rectangle} shape;

shape s;

In an enumeration type, the named value will act the same as a constant. By default, they can be considered as type "int." They can be copied from and to variables of the enumeration type. Comparison can also be made between each other. Due to the strongly typed nature of enumeration, a numeric value cannot be copied into a variable of enumeration type, but usage of type_cast help.

e.g., s = 4;   //error

s = shape'(2);   //Casting allowed

Since the default type of an enumeration is "int" so in an expression, enumeration value can be compared with an integer.

Example 4: Usage of typedef in creating enumerated datatype for FSM
In writing a code of the circuit designed using FSM, states need to be defined by means of symbols, normally alphabets. These symbols are mapped with the appropriate number of bits in the assignment process as binary-representation is necessary for hardware realization. In Verilog, this enumeration is done by using the "localparam" construct.

   e.g., considering an FSM with three states A, B, C. The code will be like:

localparam [1:0] A = 2'b00,

               B = 2'b01,

               C = 2'b10;

Declaration of signal will be written as:

reg [1:0] present_state, next_state;

The above code in Verilog can be simplified by using "typedef" and "enum" of SystemVerilog which can explicitly list the symbolic values of a set. The code using SystemVerilog will be:

typedef enum {A, B, C} State;//defining an enumerate data-type State

State present_state, next_state;//declaration of signal (State type)

The above statements are also clearer and more descriptive.

11.3 Always_type

In addition to the "general-purpose" *always* block, SystemVerilog introduces three additional procedural blocks to describe the nature of the intended hardware:

(i) *always_comb*: Usage of _comb after *always* indicates that the circuit for which this *always* block is written is purely combinational. In Verilog, *always* @ (*) is similar to *always_comb* but the only difference is in the sensitivity list. In *always* @ (*) all signals are included in the sensitivity list but in *always_comb*, the signals present in the right-hand side of the statements are only included in the sensitivity list. A warning will be issued by the system if any type of looping is observed as a violation if pure combinational circuit occurs.

(ii) *always_ff*: Usage of _ff after *always* indicates that the circuit for which this *always* block is written contains registers. The working of *always_ff* is quite similar to the normal *always* block used in Verilog. By using *always_ff* information is provided to the logic synthesis that a F/F or register is used in the statements. A warning will be issued by the system if no F/F or register is found.

(iii) *always_latch*: Use of _latch after *always* indicates that the circuit for which this *always* block is written contains a latch. A latch here means that one or more combinational loop is available in the circuit due to which an internal memory is created.

Example 5: Difference in types of *always* block.

Let's take an example of *always_type* by writing a system Verilog code for Figure 11.1.

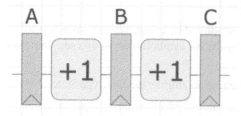

Figure 11.1

logic A_next, B_next, C_next;

logic A_r, B_r, C_r;

always_ff @(posedge clk)

begin

A_r ≤ A_next;

B_r ≤ B_next;

C_r ≤ C_next;

end

always_comb

begin

B_next = A_r + 1;

C_next = B_r + 1;

end

11.4 $log2c() Function

A very interesting feature is present in SystemVerilog which proves to be very important as the binary number system is the base of any digital circuit. This feature is a new system function, "$log2c()," which performs the same functionality as $\lceil log\, 2\, x \rceil$. This function is able to determine the required number of bits for the representation of any value.

For example, when designing a mod-M counter, the required number of bits is determined by "$log_2 M$." The code in SystemVerilog using this new feature will be:

```
module modcounter
#(parameter M = 100)      //mod-M
(input logic clock,
output logic [$clog2(M)-1:0] Q);
logic [$clog2(M)-1:0] present_state, next_state;      //signal declaration
always_ff @(posedge clk)            //register
    present_state ≤ next_state;
assign next_state = (present_state = (M-1))  //next-state logic
? 0: present_state + 1;
assign Q = present_state;            //output logic
endmodule
```

11.5 System-Verilog as a Verification Language

SystemVerilog is an object-oriented language incorporated with the benefits of object-oriented programming, inheritance, and polymorphism widely used in developing verification code by other high-level languages such as C++. These are very important features which make SystemVerilog language an HDVL (hardware description and verification language) but these are beyond the scope of the book.

Review Questions

Q1 Explain the difference between always_comb and always@(*).

Q2 Using enum datatype write a code to control traffic lights.

Q3 Design 512MB RAM using packed array and make it accessible using 2-dimentional ports.

Q4 Compare usage of packed and unpacked arrays in detail.

Q5 Compare Verilog and SystemVerilog in terms of available datatypes.

Multiple Choice Questions

Q1 The advantage of keyword "logic" as a data type is that it can be declared in place:
- **A** wire
- **B** reg
- **C** both wire and reg
- **D** None of the above

Q2 By using the facility of cast:
- **A** upacked array can be converted in packed array
- **B** packed array can be converted in unpacked array
- **C** upacked array can be converted in packed array and vice versa
- **D** None of the above

Q3 The keyword used in data-type for symbolic representation of data (values by names) is:
- **A** Symb
- **B** enum
- **C** dname
- **D** None of the above

Q4 The always block describing the circuit which contains registers is written as:
- **A** always_comb
- **B** always_latch
- **C** always_ff
- **D** None of the above

Q5 The value evaluated by $log2c(8) will be
- **A** 2
- **B** 3
- **C** 4
- **D** 6

Reference

[1] IEEE. (2019). P1800 - Standard for SystemVerilog. https://standards.ieee.org/project/1800.html (accessed 22 June 2021).

Index

Digital VLSI Design and Simulation with Verilog, First Edition. Suman Lata Tripathi,
Sobhit Saxena, Sanjeet Kumar Sinha, and Govind Singh Patel
© 2022 John Wiley & Sons Ltd. Published 2022 by John Wiley & Sons Ltd.

Printed and bound by CPI Group (UK) Ltd, Croydon, CR0 4YY

16/04/2025

14658598-0001